for AQA

A2 CHEMISTRY

Revision Guide

Sandra Clinton
Emma Poole

)XFORD
NIVERSITY PRESS

reat Clarendon Street, Oxford OX2 6DP
xford University Press is a department of the University of Oxford.
 furthers the University's objective of excellence in research, scholarship,
ıd education by publishing worldwide in
xford New York
ıckland Cape Town Dar es Salaam Hong Kong Karachi
ıala Lumpur Madrid Melbourne Mexico City Nairobi
ew Delhi Shanghai Taipei Toronto
'ith offices in
gentina Austria Brazil Chile Czech Republic France Greece
uatemala Hungary Italy Japan South Korea Poland Portugal
ngapore Switzerland Thailand Turkey Ukraine Vietnam
xford is a registered trade mark of Oxford University Press
 the UK and in certain other countries

rst published 2009

British Library Cataloguing in Publication Data
Data available
ISBN 9780199152773
10 9 8 7 6 5 4 3 2

Printed in Great Britain by Bell and Bain Ltd., Glasgow

MIX
Paper from responsible sources
FSC www.fsc.org FSC® C007785

OXFORD
UNIVERSITY PRESS

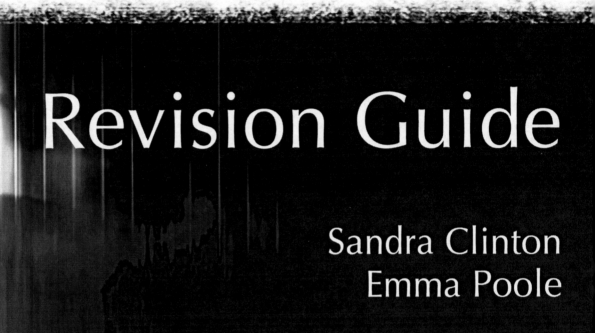

Contents

WELCOME/INTRODUCTION

Welcome to the Revision Guide for AQA A2 Chemistry. We've tried to package the course in such a way as to help you more easily go into the examination room with added confidence for success.

You'll find details on:

- how AQA assess you through the examinations together with how examinations work
- there's all the content from the course presented in Specification order (with a handy list of Specification references for easy location)
- revision guidance to help you with your planning

There are practice exam-style questions together with quick-check style questions on each spread; and new to A2 – synoptic questions!

Sandra Clinton

Emma Poole

2009

ASSESSMENT IN AQA A2 CHEMISTRY

Here are the nuts and bolts of the course: the practicalities of how AQA assesses the progress you have made during your studies.

To make the most of your exams it's essential you know how the various sections of the exams are constructed. If you know *how* the exams work you will be better placed to gain maximum marks.

Assessment overall

Your A2 Chemistry qualification is made up from three Units. Units 4 and 5 are assessed by written examination; Unit 6 is an assessment of the investigative and practical skills you have gained during the course.

Unit	Name	Length of exam	% age (total) A level marks
4	*Kinetics, Equilibria, and Organic chemistry*	1 hour 45 minutes	20%
	6–8 short answer questions plus 2–3 structured questions. Some of the questions will have synoptic elements		
5	*Energetics, Redox, and Inorgainc chemistry*	1 hour 45 minutes	20%
	6–8 short answer questions plus 2–3 longer structured questions. Some of the questions will have synoptic elements		
6	*Investigative and practical skills in chemistry* (internal assessment)		10%

Each of these three Units will examine your ability to meet the assessment objectives set out below. Work through the statements and highlight the key words.

- note that ***these are skills***, not lists of content (which are found in the Specification)

Assessment objectives plus...

Assessment objectives AO1 and AO2 are assessed across all Units while AO3 is assessed mainly in Unit 6. In essence the assessment objectives AO1 and AO2 require you to know lots of stuff – facts, figures, and chemistry; whilst AO3 expects you to know *how* chemistry works – or what makes it tick – it's all about being a practical scientist, a chemist!

Assessment objective AO1

Knowledge and understanding of science and of *How Science Works*

You should be able to:

- recognise, recall, and show understanding of scientific knowledge
- select, organise, and communicate relevant information in a variety of forms

Assessment objective AO2

Application of knowledge and understanding of science and of *How Science Works*

You should be able to:

- analyse and evaluate scientific knowledge and processes
- apply scientific knowledge and processes to unfamiliar situations, including those related to issues
- assess the validity, reliability, and credibility of scientific information

Assessment objective AO3

You should be able to:

- demonstrate and describe ethical, safe and skilful practical techniques and processes, selecting appropriate qualitative and quantitative methods
- make, record and communicate reliable and valid observations and measurements with appropriate precision and accuracy
- analyse, interpret, explain and evaluate the methodology, results and impact of their own and others' experimental and investigative activities in a variety of ways

Quality of written communication

It's all very well knowing loads of stuff but you need to be able to communicate it and get your ideas across to the examiner. You might not think it but the examiner has your interests at the centre of their job – you need to give them the easiest route to maximizing your marks.

You should:

- ensure that text is legible and that spelling, punctuation, and grammar are accurate so that meaning is clear
- select and use a form and style of writing appropriate to purpose and to complex subject matter
- organise information clearly and coherently, using specialist vocabulary where appropriate

Quality of written communication is assessed across all externally assessed Units; if you write clear, well explained answers then you should obtain any marks assigned to it.

Investigative and practical skills

Your school will decide if your work is to be internally or externally marked – this has no impact on the work you will complete but does make the system appear slightly more complicated.

There are two routes: Route T and Route X.

***Route T*: marked inTernally by your Teachers.** You will complete two components:

- Practical **S**kills **A**ssessment (PSA)
- Investigative **S**kills **A**ssignment (ISA)

***Route X*: marked eXternally by the eXam board:** You will complete two components:

- Practical **S**kills **V**erification (PSV)
- Externally **M**arked **P**ractical **A**ssignment (EMPA)

You will be prepared for these assessments during the course of your normal lessons. They will examine your ability to meet assessment objective AO3.

HOW EXAMINATIONS WORK

By now you should know about this, but read on …

Part of your course involves understanding/knowing *how science works*. Well, for maximum marks in your exam you need to know *how examinations work*. And in much the same way as science there are in-built rules that form the foundation of how examinations are constructed.

Get and speak the *lingo*: know exam-speak, play the game. Here are some popular terms which are often used in exam questions. Make sure you know what each of these terms means. For a term that requires a written answer it is most unlikely that one/two words will do!

- **Calculate**: means calculate and write down the numerical answer to the question. Remember to include your working and the units.
- **Define**: write down what a chemical/term means. Remember to include any conditions involved.
- **Describe**: write down using words and, where appropriate diagrams, all the key points. Think about the number of marks that are available when you write your answer.
- **Discuss**: write down details of the points in the given topic.
- **Explain**: write down a supporting argument using your chemical knowledge. Think about the number of marks that are available when you write your answer.
- **List**: write down a number of points. Think about the number of points required. Remember, incorrect answers will cost you marks.
- **Sketch**: when this term is used a simple freehand answer is acceptable. Remember to make sure that you include any important labels.
- **State**: write down the answer. Remember a short answer rather than a long explanation is required.
- **Suggest**: use your chemical knowledge to answer the question. This term is used when there is more than one possible answer or when the question involves an unfamiliar context.

GETTING DOWN TO REVISION

… And you should know about this too!

OK, you're committed to preparing for your examination! How do *you* go about it? Remember, there are almost as many ways to revise as there are students revising. Underneath the methods of revising there are some common goals that the revising has to achieve.

1 Boost your confidence

Careful revision will enable you to perform at your best in your examinations. So give yourself the easiest route through the work. Organise the work into small, manageable chunks and set it out in a timetable. Then each time you finish a chunk you can say to yourself 'done it', and then move on to the next one. **And give yourself a reward too!** It's amazing how much of a lift it gives you by working in this way.

2 Be successful

To be successful in A2 level chemistry you must be able to:

- recall information
- apply your knowledge to new and unfamiliar situations
- carry out precise and accurate experimental work*
- interpret and analyse both your own experimental data and that of others*

*experience gained with practical work will help you with answering questions in the examination so don't set aside all this valuable knowledge and understanding

How this revision guide can help

This book will provide you with the facts which you need to recall and some examination practice.

Use it as a working book and colour it in! Start with the Unit guides/maps that show you each of the topics covered. Read them and highlight the areas which you are already confident about. Then, in a different colour mark off the sections one at a time as your revision progresses. The more colour on your map the more work you have done to prepare for your examination. By doing this you will feel positive about what you have achieved rather than negative about what you still have to do.

For your revision programme you might like to use some or all of the following strategies:

- read through the topics one at a time and try the quick questions
- choose a topic and make your own condensed summary notes
- colour important diagrams
- highlight key definitions or write them onto flash cards
- work through facts until you can recall them
- constantly test your recall by covering up sections and writing them from memory
- ask your friends and family to test your recall
- make posters for your bedroom walls
- use the 'objectives' as a self-test list
- carry out exam practice
- work carefully through the material on each page, highlighting the guide/map sections as you go
- make 'to do' lists like those on the exam practice pages
- and don't forget about HSW (How Science Works)!

Whatever strategies you use, measure your revision in terms of the progress you are making rather than the length of time you have spent working. You will feel much more positive if you are able to say specific things you have achieved at the end of a day's revision rather than thinking, 'I spent eight hours inside on a sunny day!'. Don't sit for extended periods of time. Plan your day so that you have regular breaks, fresh air, and things to look forward too.

Watch out: revision is an active occupation - just reading information is not enough! You will need to be active in your work for your revision to be successful.

Improving your recall: A good strategy for recalling information is to focus on a small number of facts for five minutes. Copy out the facts repeatedly in silence for five minutes then turn your piece of paper over and write them from memory. If you get any wrong then just write these out for five minutes. Finally test your recall of all the facts. Come back to the same facts later in the day and test your self again. Then revisit them the next day and again later in the week. By carrying out this process they will become part of your long term memory – you will have learnt them!

Past paper practice: Once you have built up a solid factual knowledge base you need to test it by completing past paper practice. It might be a good idea to tackle several questions on the same topic from a number of papers rather than working through a whole paper at once. This will enable you to identify any weak areas so that you can work on them in more detail. Finally, remember to complete some mock exam papers under exam conditions.

A final word (or two) for the examination room

Unlike the GCSE examination this examination does not have a foundation tier and a higher tier so you must be prepared to answer questions on all the topics outlined in the Specification. Here are some obvious, and not so obvious thoughts:

- read through all the questions (*obvious*)
- identify which questions you can answer well (*obvious*)
  - start by answering these questions (*not so obvious*)
- once you have read a question carefully make sure you answer that question and NOT something you might think is the question (*not so obvious*)
- look at the number of marks that are available for each question and take this into account when you write your answer (*obvious and not so obvious*)
- answer space: the amount of space left for the answer will give you an indication of the length of answer the examiner expects (*obvious*)
  - short space = short answer (*obvious*)
  - longer space(1) = extended answer probably required; perhaps a sentence, or two; a calculation with working; a list containing a selection items (*obvious*)
  - longer space(2) = one word answer unlikely to be sufficient (*not so obvious*)
- answer all the questions, even if you have to guess at some (*obvious*)
- pace yourself and try to leave enough time to check your answers at the end (*obvious*)

4.01 Kinetics 1

OBJECTIVES

By the end of this section you should know:

- *the factors that affect the rate of a reaction*
- *how to analyse the results of experiments to reveal the order of reaction with respect to a reactant*
- *how to derive the rate equation*
- *how to use the rate equation*

The rate of reaction

The rate of reaction is a way of measuring how quickly a reactant is being used up or how quickly a product is being made. When we measure the rate of reaction we measure the change in concentration of a species over a period of time and the units used are mol dm^{-3} s^{-1}

Factors that affect the rate of reaction

- **Temperature** – As the temperature increases the particles gain more kinetic energy. This means they collide more often and when they do collide more particles have enough energy to react (activation energy) so the rate of reaction increases.
- **Catalysts** – A catalyst increases the rate of reaction by offering an alternative reaction pathway with a lower activation energy. This means that more particles have energy equal to, or greater than, the activation energy, so more of the collisions are successful and the rate of reaction increases.
- **Concentration** – As the concentration of solutions of reactants increases the particles become closer together. This means that the particles collide more often so the rate of reaction increases.
- **Pressure** – As the pressure of gaseous reactants increases the particles are closer together. This means that the particles collide more often so the rate of reaction increases.
- **Surface area** – Increasing the surface area of a solid reactant increases the rate of reaction. As the surface area increases the collision frequency increases so the rate of reaction goes up.

The rate equation

The **rate equation** is written in the form:

$$\text{rate} = k[A]^m[B]^n$$

- Notice the square brackets which indicate that we are measuring concentrations in units of mol dm^{-3}.
- [A] is the concentration of species A.
- [B] is the concentration of species B.
- k is the rate constant, which can have different units
- m is the order of reaction with respect to A.
- n is the order of reaction with respect to B.

The rate equation cannot be worked out by looking at the stoichiometric equation. In fact, there may be species that appear on the stoichiometric equation that do not appear in the rate equation and there may be species, such as a catalyst, that appear in the rate equation but do not appear in the stoichiometric equation. The rate equation can only be determined by experimental methods.

The overall order of reaction

The overall order of the reaction is the sum of the orders of reaction with respect to the individual species. In this example:

$$\text{rate} = k[A]^2[B]$$

The reaction is second order with respect to A and first order with respect to B. The overall order of reaction is given by 2+1, so the overall order of this reaction is 3.

Orders of reaction

Zero order

If the order of reaction with respect to a species is zero then changing the concentration of that species has no effect on the rate of reaction. The species is not involved in the **rate determining step**.

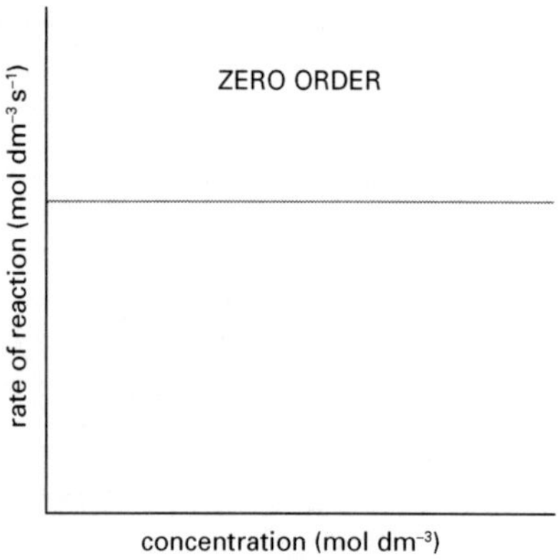

If a reaction is second order with respect to A and zero order with respect to B. The rate equation is:

$$\text{rate} = k[A]^2[B]^0$$

Since anything to the power of zero is equal to 1 this equation simplifies to:

$$\text{rate} = k[A]^2$$

First order

If a reaction is first order with respect to a species then the concentration of that species has a direct effect on the rate of reaction. If the concentration of the species is doubled the rate of reaction will also be doubled. If the concentration of the species is tripled then the rate of reaction is also tripled.

Note that the line has a positive gradient and that it goes through the origin.

If a reaction is first order with respect to A and first order with respect to B the rate equation is:

$$\text{rate} = k[A]^1[B]^1$$

Since anything to the power of 1 remains unchanged this is simplified to:

$$\text{rate} = k[A][B]$$

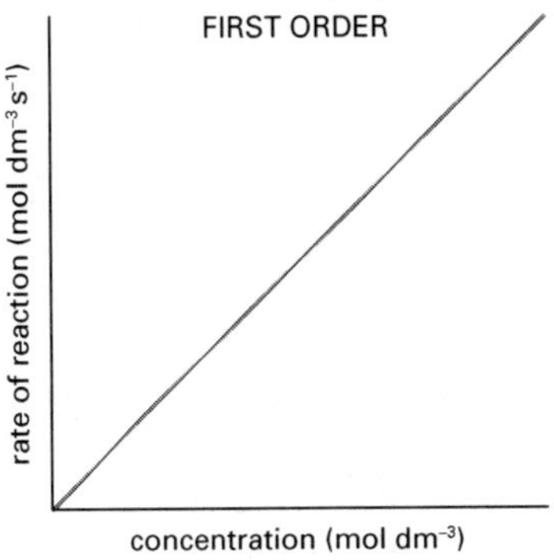

Second order

If a reaction is found to be second order with respect to a species then the concentration of that species also has a direct effect on the rate of reaction.

If the concentration of the species is doubled the rate of reaction increases four times ($2^2 = 4$).

If the concentration of the species is tripled the rate of reaction increases nine times ($3^2 = 9$).

Note that the line curves upwards and that it goes through the origin.

If a reaction is second order with respect to A the rate equation is:

$$\text{rate} = k[A]^2$$

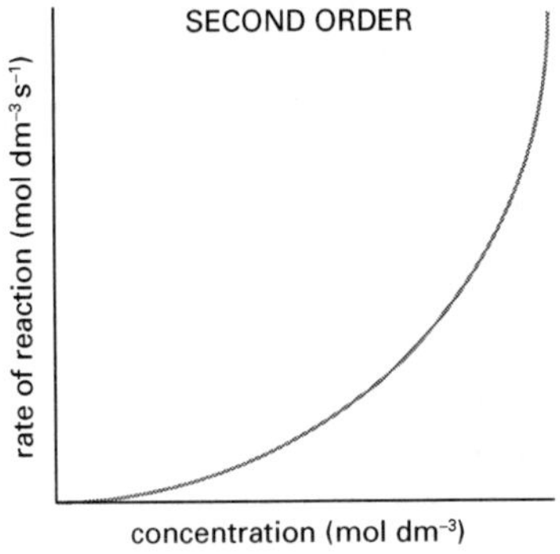

Finding the order of reaction

The order of reaction for a species can be worked out using **the initial rates method**.

In this method the experiment is carried out several times using different concentrations of the reactants and any other species that might affect the rate of reaction, such as a catalyst.

The concentrations of all species except one are kept constant and the initial rate of reaction is measured with different values of the concentration of that one species.

Any changes in the initial rate of reaction must have been caused by the change in the concentration of that particular species. This information can be used to calculate the order of reaction with respect to that species.

The experiment is then repeated, with the concentration of the next species to be investigated being changed.

Questions

1 The rate equation for a reaction is

Rate = $k[A]^2[B]^2$

What is the overall order of reaction?

2 The rate equation for a reaction is

Rate = $k[A][B]$

What are the units of the rate constant k?

3 A reaction is first order with respect to A and zero order with respect to B. What is the rate equation for this reaction?

Worked example

Iodine reacts with propanone in the presence of an acid catalyst. From the data in the table work out the order of the reaction with respect to propanone, iodine, and hydrogen ions. Write the rate equation for the reaction.

$$CH_3COCH_3 + I_2 \xrightarrow{H^+} CH_3COCH_2I + HI$$

concentration of CH_3COCH_3 (mol dm^{-3})	concentration of I_2 (mol dm^{-3})	concentration of H^+ (mol dm^{-3})	initial rate of reaction (mol dm^{-3} s^{-1})
0.5	0.5	0.5	1
0.5	1.0	0.5	1
1.0	0.5	0.5	2
0.5	0.5	1.0	2

Answer

The reaction is first order with respect to CH_3COCH_3

The reaction is zero order with respect to I_2

The reaction is first order with respect to H^+

So the rate equation is

Rate = $k[CH_3COCH_3][H^+]$

4.02 Kinetics 2

OBJECTIVES

By the end of this section you should:

- *be able to explain the quantitative effects in temperature on the rate constant, k*
- *know that the orders of reaction with respect to reactants can be used to provide information about the rate determining step*
- *know about some of the methods used to carry out kinetic studies into the order of reaction*

The rate constant and temperature

- An increase in temperature increases the rate of reaction.
- The rate equation is:

$$\text{rate} = k[\text{A}]^m[\text{B}]^n$$

Changing the temperature does not affect the concentrations of A or B. Therefore the rate constant k must increase as the temperature increases. This is true for exothermic and for endothermic reactions.

As a result it is very important that if we are investigating the effect of changing the concentration of A or B on the rate of reaction that we keep the temperature constant.

- A large value of k means a fast rate of reaction.
- A small value of k means a slow rate of reaction.
- A catalyst provides an alternative reaction pathway with a lower activation energy so it increases the value of the rate constant, k.

Finding the order of reaction

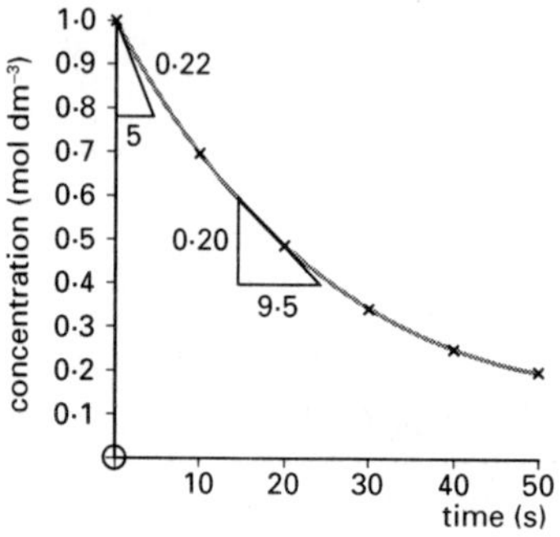

The order of reaction for a species can also be worked out using the graphical method.

In this method the experiment is carried out once and the concentration of the reactant is measured at different times. These results are used to draw a concentration–time graph.

Tangents are drawn and the gradient of these tangents gives the rate of reaction.

The gradient at the initial concentration of 1.0 mol dm^{-3} is given by 0.22 mol dm^{-3}/5 s = 0.044 mol dm^{-3} s^{-1}

The gradient is then worked out at another point in the graph, normally at half the initial concentration.

The gradient at half the initial concentration (0.5 mol dm^{-3}) is given by 0.20 mol dm^{-3}/9.5 s = 0.021 mol dm^{-3} s^{-1}

The second rate of reaction is approximately half of the value of the initial rate of reaction.

As the rate of reaction has halved as the concentration has halved, the reaction is first order.

Using half-lives

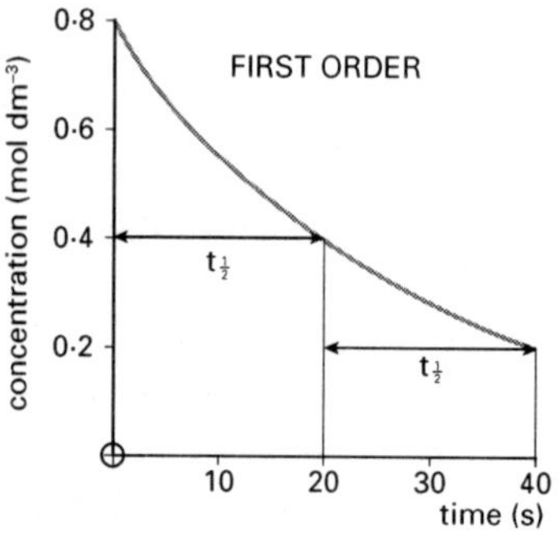

The half-life of a reaction is the time it takes for the concentration of a reactant to halve.

The half-life of a reaction can be determined from a concentration–time graph.

First order reactions

First order reactions have a constant half-life.

It takes 20 seconds for the concentration to halve from 0.8 mol dm^{-3} to 0.4 mol dm^{-3}.

It takes a further 20 seconds for the concentration to halve from 0.4 mol dm^{-3} to 0.2 mol dm^{-3}.

As the half-life remains constant this means that this is a first order reaction.

Zero order reactions

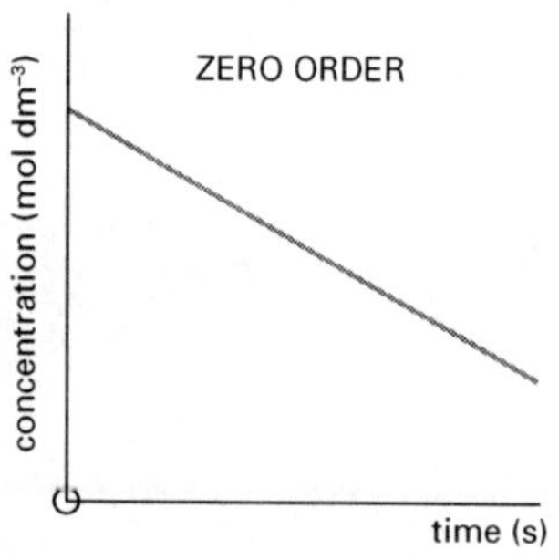

By contrast this is the concentration – time graph for a zero order reaction

The gradient of the line is constant. This means that the rate of reaction is constant. It shows that the amount of the reactant has no effect on the rate of reaction.

Second order reactions

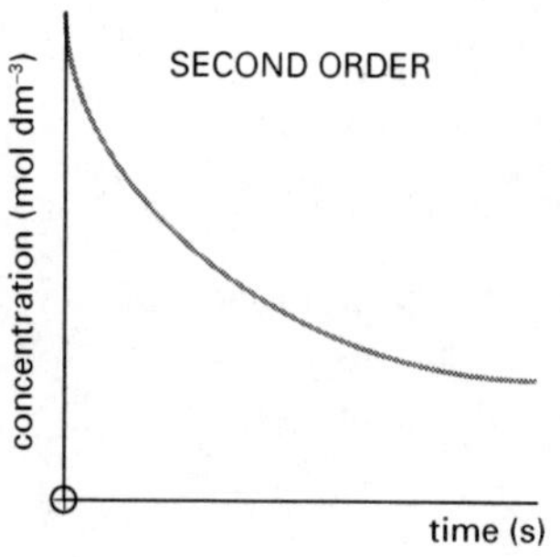

For a second order reaction the half-life increases as the concentration decreases.

Rate equations and mechanisms

Although some reactions involve just one step many reactions involve a number of steps that happen in order. The slowest step in the reaction is called the **rate determining step** (or rate limiting step). This step has the highest activation energy.

Chemists are very interested in the mechanisms of a reaction. If they know which bonds are being broken and which bonds are being made and the order in which this is happening it helps them to understand how the new compound is made and helps them to design useful new compounds.

The rate equation gives us information about the mechanisms because it tells us the species that are involved in the rate determining step.

The order of reaction with respect to a species tells us the number of molecules of that species in the rate determining step.

For example if the rate equation for a reaction is:

$$rate = k[A]^2[B]$$

It tells us the reaction is second order with respect to A and first order with respect to B.

This means that the rate determining step involves two molecules of A and one molecule of B or something that is made from two molecules of A and one molecule of B.

The nucleophilic substitution of haloalkanes

A haloalkane will react with an aqueous solution of sodium hydroxide to form an alcohol and sodium halide.

$$
\begin{array}{c}
CH_3 \\
| \\
CH_3-C-Br \\
| \\
H
\end{array}
+ NaOH \longrightarrow
\begin{array}{c}
CH_3 \\
| \\
CH_3-C-OH \\
| \\
H
\end{array}
+ NaBr
$$

2-bromopropane + sodium hydroxide ⟶ propan-2-ol + sodium bromide

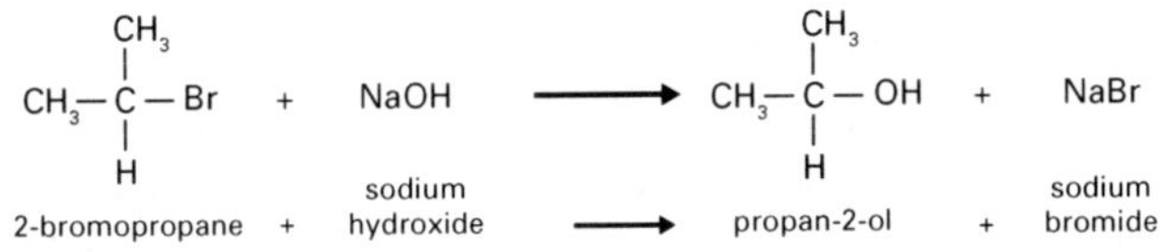

This is an example of a nucleophilic substitution reaction. The nucleophile is the hydroxide, OH^- ion.

Two possible mechanisms are proposed for the reaction.

Mechanism 1

step 1
$$
\begin{array}{c}
CH_3 \\
| \\
CH_3-C-Br \\
| \\
H
\end{array}
\xrightarrow{\text{slow}}
\begin{array}{c}
CH_3 \\
| \\
CH_3-C^+ \\
| \\
H
\end{array}
: Br^-
$$

step 2
$$
\begin{array}{c}
CH_3 \\
| \\
CH_3-C^+ \\
| \\
H
\end{array}
:OH^-
\xrightarrow{\text{fast}}
\begin{array}{c}
CH_3 \\
| \\
CH_3-C-OH \\
| \\
H
\end{array}
$$

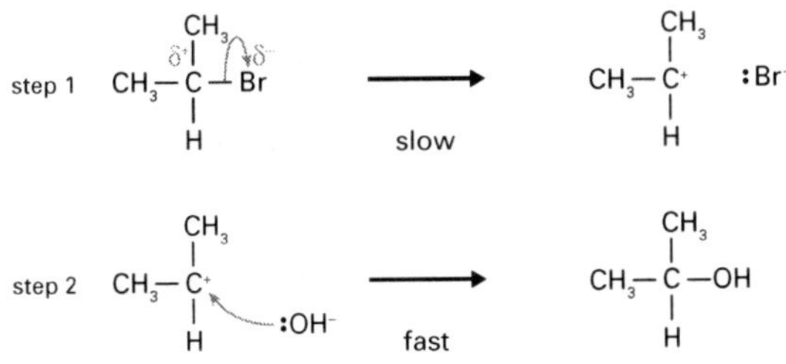

The rate determining step is the slowest step in the reaction. Here the slowest step is the first one which only involves the haloalkane, which is 2-bromopropane.

Mechanism 2

$$
\begin{array}{c}
CH_3 \\
| \\
CH_3-C-Br \\
| \\
H
\end{array}
:OH^-
\longrightarrow
\left[
\begin{array}{c}
CH_3 \\
| \quad \text{Br} \\
CH_3-C \\
| \quad \text{OH} \\
H
\end{array}
\right]^-
\longrightarrow
\begin{array}{c}
CH_3 \\
| \\
CH_3-C-OH \\
| \\
H
\end{array}
+ :Br^-
$$

transition state

slow

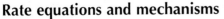

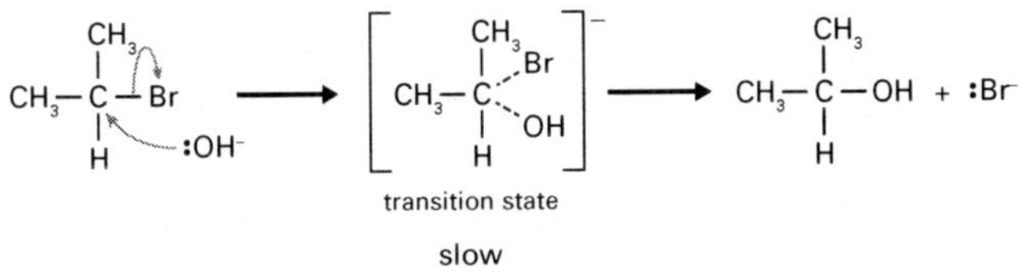

The reaction takes place in one step via a transition state in which the bond between the carbon atom and the bromine atom is partially broken and the bond between the carbon atom and the oxygen is partially made. Both the haloalkane and the hydroxide ion are involved in the rate determining step.

The rate equation for the reaction is found to be:

$$rate = k[C_3H_7Br][OH^-]$$

A suggested mechanism must be consistent with the orders of reaction shown in the rate equation.

As both the haloalkane and the hydroxide ion appear in the rate equation then the reaction could proceed via mechanism 2 which involves both species in the rate determining step but could not proceed via mechanism 1 because the hydroxide, which is in the rate equation, is not involved in the rate determining (slowest) step of the mechanism.

Worked example

In a reaction A reacts with B in the presence of a catalyst, C, to produce G.

$$A + B \xrightarrow{\text{catalyst, C}} G$$

Two possible mechanisms are proposed for the reaction

Mechanism 1   **Step 1** $A + C \xrightarrow{\text{slow}} D$

               Step 2 $D \xrightarrow{\text{fast}} E + C$

               Step 3 $E + B \xrightarrow{\text{fast}} G$

Mechanism 2   **Step 1** $A + B + C \xrightarrow{\text{slow}} G$

The rate equation is found to be: rate = $k[A][C]$

Which, if either, of these mechanisms is consistent with the rate equation?

In mechanism 1 the rate determining step involves reactant A and the catalyst C. As these both appear in the rate equation, mechanism 1 is consistent with the rate equation.

However, in mechanism 2 the rate determining step involves both the reactants A and B plus the catalyst C. As B does not appear in the rate determining step this mechanism is not consistent with the rate equation.

Methods of measuring the rate of a reaction

Methods of following the rate of reaction include:

- titration
- colorimetry
- collecting the volume of a gas produced

Questions

1 What is the half-life of a reaction?

2 How can a concentration–time graph be used to prove that a reaction is first order?

3 What does a high value for the rate constant k indicate?

4.03 Equilibria 1

OBJECTIVES

By the end of this section you should:

- *know that many reactions are reversible*
- *know how to use Le Chatelier's principle to explain the effects of changes of temperature, pressure, and concentration*
- *know how to predict the effect of changing temperature on the position of an equilibrium*
- *know why compromise temperatures and pressures may be used in industrial processes*
- *know that K_c is the equilibrium constant and that it can be calculated from equilibrium concentrations for systems at constant pressure*
- *be able to construct expressions for K_c*
- *be able to predict the effects of changes of temperature on the value of the equilibrium constant*
- *know that the equilibrium constant is not affected by changes in concentration or by the presence of a catalyst*

Le Chatelier's Principle

Many chemical reactions are reversible. When a reversible reaction reaches a dynamic equilibrium the rate of the forward reaction is equal to the rate of the backward reaction. If the temperature is changed the position of equilibrium moves. We can apply Le Chatelier's Principle to work out how the position of equilibrium will change. If the temperature is increased the equilibrium moves in the endothermic reaction. In a reaction which is exothermic in the forward direction, increasing the temperature moves the equilibrium towards the endothermic side and reduces the yield of the product made.

Le Chatelier's Principle

Le Chatelier's Principle states that if a small change is applied to a system in dynamic equilibrium, the position of the equilibrium moves in such a way as to minimize the effect of the change

Worked example

Explain the term 'dynamic equilibrium'.

In a dynamic equilibrium the rate of the forward reaction is equal to the rate of the backward reaction so that there is no change in the concentrations of the products and reactants.

Concentration

Concentration is equal to the number of moles of a substance in 1 dm^3 of a solution. It is measured in units of mol dm^{-3}

[A] means the concentration of substance A in units of mol dm^{-3}. Notice how square brackets are used.

Worked example

At equilibrium a mixture was found to contain 0.001 moles of A and 0.02 moles of B. The total volume was 2.0 dm^3. What are the concentrations of A and B?

$$\text{Concentration of A} = \frac{0{\cdot}001 \text{ moles}}{2 \text{ dm}^3} = 0{\cdot}0005 \text{ mol dm}^{-3}$$

$$\text{Concentration of B} = \frac{0{\cdot}02 \text{ moles}}{2 \text{ dm}^3} = 0{\cdot}01 \text{ mol dm}^{-3}$$

Worked example

The following system is at equilibrium.

$X + Y \rightleftharpoons Z$

At equilibrium there was found to be 0.04 moles of X, 0.02 moles of Y, and 0.005 moles of Z. The total volume is 500 cm^3. What are the concentrations of X, Y, and Z?

500 cm^3 is equal to 0.5 dm^3.

$$\text{Concentration of X} = \frac{0{\cdot}04 \text{ moles}}{0{\cdot}5 \text{ dm}^3} = 0{\cdot}08 \text{ mol dm}^{-3}$$

$$\text{Concentration of Y} = \frac{0{\cdot}02 \text{ moles}}{0{\cdot}5 \text{ dm}^3} = 0{\cdot}04 \text{ mol dm}^{-3}$$

$$\text{Concentration of Z} = \frac{0{\cdot}005 \text{ moles}}{0{\cdot}5 \text{ dm}^3} = 0{\cdot}01 \text{ mol dm}^{-3}$$

The equilibrium constant

- The equilibrium constant, K_c, has a fixed value for any particular equilibrium reaction at a given temperature.
- It is a measure of how far a reaction has proceeded but gives no information about the rate of the reaction.
- A high value of K_c indicates that the equilibrium favours the products.
- A low value of K_c indicates that the equilibrium favours the reactants.

Finding the equilibrium constant

The equilibrium constant is written in terms of the equilibrium concentrations. For the reaction:

$$aA + bB \rightleftharpoons cC + dD$$

$$K_c = \frac{[C]^c\,[D]^d}{[A]^a\,[B]^b}$$

- All the concentration must be equilibrium concentrations.
- If the system has yet to reach equilibrium this equation will not apply. The system will continue to react until equilibrium is established.
- The equilibrium concentration of each term is raised to the power of the number used in the balanced equation.
- The equilibrium concentrations of the products are written on the top of the fraction.
- The equilibrium concentrations of the reactants are written on the bottom of the fraction.
- K_c is a concentration term, so it can only be applied to gases and solutions.

Worked example

Write the K_c expression for this reaction.

$N_2(g) + 3H_2(g) \rightleftharpoons 2NH_3(g)$

$$K_c = \frac{[NH_3]^2}{[N_2]\,[H_2]^3}$$

The products are written on the top of the fraction.

The reactants are written on the bottom of the fraction.

The equilibrium concentration of ammonia is raised to the power of 2.

The equilibrium concentration of hydrogen is raised to the power of 3.

Homogeneous equilibria

In homogeneous equilibria all the species are in the same phase. For example they are

- all gases
- all solutions

Working out the units of K_c

$H_2(g) + Cl_2(g) \rightleftharpoons 2HCl(g)$

$$K_c = \frac{[HCl]^2}{[H_2]\,[Cl_2]}$$

$$= \frac{\cancel{(\text{mol dm}^{-3})}\ \cancel{(\text{mol dm}^{-3})}}{\cancel{(\text{mol dm}^{-3})}\ \cancel{(\text{mol dm}^{-3})}}$$

For this equation the units cancel so there are no units for K_c, and this is written as 'no units'. However, this is not always the case and the units must be worked out each time.

Worked example

Dinitrogen tetroxide, $N_2O_4(g)$, decomposes to form nitrogen dioxide, $NO_2(g)$.

$N_2O_4(g) \rightleftharpoons 2NO_2(g)$

Give the expression for K_c for this reaction. Include the units.

$$K_c = \frac{[NO_2]^2}{[N_2O_4]}$$

$$= \frac{\cancel{(\text{mol dm}^{-3})} \times (\text{mol dm}^{-3})}{\cancel{(\text{mol dm}^{-3})}}$$

$$= \text{mol dm}^{-3}$$

Factors that affect the value of K_c

A catalyst will not affect the position of equilibrium so it will not affect the value of K_c. It will simply mean that we reach the position of equilibrium more quickly.

For solutions, if the concentration of one of the substances is changed then the system will no longer be in equilibrium so it will adjust until it re-establishes equilibrium. At that point the value of K_c will not have changed.

For gases, if the pressure is changed and the number of molecules on each side of the balanced equation is not the same then the system will no longer be in equilibrium. It will adjust until it re-establishes equilibrium, at which point the value of K_c will not have changed.

However **temperature** will affect the position of equilibrium. If the reaction is exothermic in the forward direction increasing the temperature will move the position of equilibrium in the endothermic direction:

- reducing the amount of products
- increasing the amount of reactants

$$K_c = \frac{[\text{products}]}{[\text{reactants}]}$$

less ([products])

more ([reactants])

Increasing the temperature of an exothermic reaction will decrease the value of K_c.

Increasing the temperature of an endothermic reaction will increase the value of K_c.

Worked example

Sulfur dioxide reacts with oxygen to form sulfur trioxide. The reaction is reversible.

$2SO_2(g) + O_2(g) \rightleftharpoons 2SO_3(g)$

Give the expression for K_c for this reaction. Include the units.

$$K_c = \frac{[SO_3]^2}{[SO_2]^2\,[O_2]}$$

$$\text{units} = \frac{\cancel{(\text{mol dm}^{-3})} \times \cancel{(\text{mol dm}^{-3})}}{\cancel{(\text{mol dm}^{-3})} \times \cancel{(\text{mol dm}^{-3})} \times (\text{mol dm}^{-3})}$$

$$= \frac{1}{\text{mol dm}^{-3}}$$

$$= \text{mol}^{-1}\ \text{dm}^3$$

Worked example

The manufacture of ammonia is based on this equilibrium.

$N_2(g) + 3H_2(g) \rightleftharpoons 2NH_3(g)$

What is the effect of adding a catalyst on:

a *the rate at which equilibrium is attained*

b *the position of equilibrium*

c *the value of K_c?*

a The rate is faster so it takes less time to reach equilibrium.

b There is no change to the position of equilibrium.

c There is no change in the value of K_c.

Questions

1 How is concentration calculated?

2 Which factor or factors will affect the value of K_c for a given reaction?

3 Write the equilibrium expression for the reaction:

$2A + 3B \rightleftharpoons 2C$

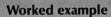

4.04 Equilibria 2

OBJECTIVES

By the end of this section you should be able to:

- *perform calculations using* K_c

The equilibrium constant, K_c

- Remember the **equilibrium constant** is not affected by changes in concentration or pressure or by the presence of a catalyst. It is only affected by changes in temperature.
- The equilibrium constant is a measure of how far the reaction will proceed but it gives no information on how fast the reaction is.
- The concentration for solids and pure liquids is constant so it is not included in the expression for K_c.

K_c for a solid

In the thermal decomposition of limestone:

$CaCO_3(s) \rightleftharpoons CaO(s) + CO_2(g)$

Calcium carbonate and calcium oxide are both solids so the expression for the rate constant is

$K_c = [CO_2]$

Calculating the value of K_c

When we know the number of moles at equilibrium

Step 1 Write the expression for K_c.

Step 2 Work out the concentrations (this is given by the number of moles divided by the volume).

Step 3 Substitute in the values.

Step 4 Calculate the answer. Remember to include the units.

Worked example

Find the value of K_c for this equilibrium:

$A + 2B \rightleftharpoons 2C$

The equilibrium mixture contained 0.3 moles of A, 0.4 moles of B, and 6.0 moles of C. The total volume was 2.0 dm³.

$$K_c = \frac{[C]^2}{[A]\,[B]^2}$$

$$\text{Concentration of A} = \frac{0{\cdot}3 \text{ mol}}{2{\cdot}0 \text{ dm}^3} = 0{\cdot}15 \text{ mol dm}^{-3}$$

$$\text{Concentration of B} = \frac{0{\cdot}4 \text{ mol}}{2{\cdot}0 \text{ dm}^3} = 0{\cdot}20 \text{ mol dm}^{-3}$$

$$\text{Concentration of C} = \frac{6{\cdot}0 \text{ mol}}{2{\cdot}0 \text{ dm}^3} = 3{\cdot}0 \text{ mol dm}^{-3}$$

$$K_c = \frac{3{\cdot}0 \text{ mol dm}^{-3} \times 3{\cdot}0 \text{ mol dm}^{-3}}{0{\cdot}15 \text{ mol dm}^{-3} \times 0{\cdot}20 \text{ mol dm}^{-3} \times 0{\cdot}20 \text{ mol dm}^{-3}}$$

$$K_c = 1500 \text{ mol}^{-1} \text{ dm}^3$$

Worked example

Find the value of K_c for this equilibrium:

$A + 2B \rightleftharpoons C + D$

The equilibrium mixture contained 0.4 moles of A, 0.3 moles of B, 0.10 moles of C, and 0.10 moles of D. The total volume was 2.5 dm³.

$$K_c = \frac{[C]\,[D]}{[A]\,[B]^2}$$

$$\text{Concentration of A} = \frac{0{\cdot}4 \text{ mol}}{2{\cdot}5 \text{ dm}^3} = 0{\cdot}16 \text{ mol dm}^{-3}$$

$$\text{Concentration of B} = \frac{0{\cdot}3 \text{ mol}}{2{\cdot}5 \text{ dm}^3} = 0{\cdot}12 \text{ mol dm}^{-3}$$

$$\text{Concentration of C and D} = \frac{0{\cdot}1 \text{ mol}}{2{\cdot}5 \text{ dm}^3} = 0{\cdot}04 \text{ mol dm}^{-3}$$

$$K_c = \frac{(0{\cdot}04 \cancel{\text{mol dm}^{-3}}) \times (0{\cdot}04 \cancel{\text{mol dm}^{-3}})}{(0{\cdot}16 \text{ mol dm}^{-3}) \times (0{\cdot}12 \cancel{\text{mol dm}^{-3}}) \times (0{\cdot}12 \cancel{\text{mol dm}^{-3}})}$$

$$K_c = \frac{0{\cdot}0016}{0{\cdot}002304} = 0{\cdot}694 \text{ mol}^{-1} \text{ dm}^3$$

When we only know the number of moles at the start of the experiment

Step 1 Write the expression for K_c.

Step 2 State the number of moles of each reactant and product at the start.

Step 3 Calculate the number of moles of each reactant and product present at equilibrium.

Step 4 Work out the concentrations (number of moles divided by the volume).

Step 5 Substitute in the values.

Step 6 Calculate the answer. Remember to include the units.

Worked example

Find the value of K_c for this equilibrium:

$A + B \rightleftharpoons C$

In an experiment 0.5 moles of A and 0.3 moles of B were mixed in a 5 dm³ container. At equilibrium there was found to be 0.3 moles of A.

$A + B \rightleftharpoons C$

$$K_c = \frac{[C]}{[A]\,[B]}$$

	number of moles		
	A	B	C
at start	0·5	0·3	0·0
at equilibrium	0·3		

In the reaction, A : B : C is 1 : 1 : 1.

The number of moles of A that have reacted is 0.2, which means that 0.2 moles of B will also have reacted leaving

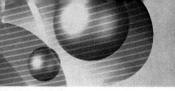

0.1 moles of B at equilibrium.

0.2 moles of C will have been made so there is 0.2 moles of C present at equilibrium.

$$A + B \rightleftharpoons C$$

$$K_c = \frac{[C]}{[A]\,[B]}$$

	A	B	C
at equilibrium	0·3 mol	0·1 mol	0·2 mol

$$\text{Concentration of A} = \frac{0{\cdot}3 \text{ mol}}{5 \text{ dm}^3} = 0{\cdot}06 \text{ mol dm}^{-3}$$

$$\text{Concentration of B} = \frac{0{\cdot}1 \text{ mol}}{5 \text{ dm}^3} = 0{\cdot}02 \text{ mol dm}^{-3}$$

$$\text{Concentration of C} = \frac{0{\cdot}2 \text{ mol}}{5 \text{ dm}^3} = 0{\cdot}04 \text{ mol dm}^{-3}$$

$$K_c = \frac{[C]}{[A]\,[B]}$$

$$K_c = \frac{0{\cdot}04 \text{ mol dm}^{-3}}{0{\cdot}06 \text{ mol dm}^{-3} \times 0{\cdot}02 \text{ mol dm}^{-3}}$$

$$= 33 \text{ mol}^{-1} \text{ dm}^3$$

Finding equilibrium amounts from K_c values

For the reaction $A \rightleftharpoons B \quad K_c = 5.0$

If 10 moles of A were allowed to reach equilibrium how many moles of B would be present in the equilibrium mixture?

At equilibrium there are x moles of B. So we can work out the number of moles of the other substances in terms of x.

number of moles		
	A	B
at start	10	0
at equilibrium	$10 - x$	x

We can find the equilibrium concentrations by dividing the number of moles by the volume, but as we do not know the volume we will call it V.

$$\text{Concentration of A} = \frac{10 - x}{V}$$

$$\text{Concentration of B} = \frac{x}{V}$$

$$K_c = \frac{[B]}{[A]} = 5{\cdot}0$$

$$5{\cdot}0 = \frac{\left(\frac{x}{V}\right)}{\left(\frac{10 - x}{V}\right)}$$

$$50 - 5x = x$$

$$50 = 6x$$

$$x = 8{\cdot}3$$

So there are 8.3 moles of B in the equilibrium mixture.

Worked example

Calculate the value of K_c for the reaction from the data given:

$A \rightleftharpoons B + C$

0.08 moles of A was placed into a 2.0 dm^3 vessel and heated. At equilibrium 75% of the A had dissociated.

If 75% of the A that was originally present has dissociated it leaves 25% or 0.02 moles of A. We can now work out the equilibrium amounts of the other species.

$$A \rightleftharpoons B + C$$

$$K_c = \frac{[B]\,[C]}{[A]}$$

number of moles			
	A	B	C
at start	0·08	0	0
at equilibrium	0·02	0·06	0·06

down by 0·06

B and C will be up by 0·06 moles

$$\text{Concentration of A} = \frac{0{\cdot}02 \text{ mol}}{2 \text{ dm}^3} = 0{\cdot}01 \text{ mol dm}^{-3}$$

$$B = \frac{0{\cdot}06 \text{ mol}}{2 \text{ dm}^3} = 0{\cdot}03 \text{ mol dm}^{-3}$$

$$C = \frac{0{\cdot}06 \text{ mol}}{2 \text{ dm}^3} = 0{\cdot}03 \text{ mol dm}^{-3}$$

$$K_c = \frac{(0{\cdot}03 \text{ mol dm}^{-3}) \times (0{\cdot}03 \text{ mol dm}^{-3})}{(0{\cdot}01 \text{ mol dm}^{-3})}$$

$$K_c = \frac{0{\cdot}0009 \text{ mol dm}^{-3}}{0{\cdot}01}$$

$$K_c = 0{\cdot}09 \text{ mol dm}^{-3}$$

Questions

1 In the reaction:

$A + 2B \rightleftharpoons C$

the equilibrium mixture contained 0.03 moles of A, 0.2 moles of B, and 0.04 moles of C. The total volume was 0.5 dm^3

a Write the expression for K_c including units.

b Find the concentration of A, B, and C at equilibrium

c Calculate the value of K_c.

2 For the equilibrium $X + 2Y \rightleftharpoons Z$

0.5 moles of X and 0.8 moles of Y were reacted together. At equilibrium there was found to be 0.3 moles of X. Calculate the number of moles of Y and Z present at equilibrium.

3 For the equilibrium $A + 2B \rightleftharpoons 2C$

0.02 moles of A and 0.10 moles of B were reacted together. At equilibrium 50% of A had reacted.

Calculate the number of moles of A, B, and C present at equilibrium.

4.05 Acids and bases 1

OBJECTIVES

By the end of this section you should:

- *know that in neutralization reactions H^+ ions react with OH^- ions to produce water*
- *know that an acid is a proton donor*
- *know that a base is a proton acceptor*
- *know that $pH = -log_{10} [H^+]$*
- *know how to calculate the pH of a solution*
- *be able to calculate the pH of a strong acid and of a strong base*
- *be familiar with the ionic product of water*
- *know how to construct an expression for K_a for a weak acid and use it to find the pH of a weak acid*
- *know that $pK_a = -log_{10}K_a$*

Brønsted–Lowry acids and bases

- An acid is a proton (H^+) donor.
- A base is a proton acceptor.
- The species formed when an acid loses an H^+ ion is the acid's **conjugate base**.
- The species formed when a base gains an H^+ is the base's **conjugate acid**.

In the equilibrium:

$$HCl + H_2O \rightleftharpoons H_3O^+ + Cl^-$$

- Cl^- is the conjugate base of HCl.
- H_3O^+ is the conjugate acid of H_2O.
- The pair is linked by the gain or loss of a H^+ ion.

Definition and determination of pH

pH is equal to –log base 10 of the concentration of hydrogen ions.

$$pH = -\log_{10}[H^+]$$

Reminder

Make sure you press the log key not the ln key on your calculator.

Strong acids like hydrochloric acid, HCl, are completely ionized in water.

$$HCl \rightarrow H^+ + Cl^-$$

So we can use the concentration of the acid to calculate the concentration of the H^+ ions and then use this to find the pH.

Worked example

Find the pH of a 0.2 mol dm^{-3} solution of HCl.

HCl is a strong acid so it is completely ionized in water.

$HCl \rightarrow H^+ + Cl^-$

0.2 0.2

$pH = -\log_{10}[H^+]$

$pH = -\log_{10}0.2$

$pH = 0.70$

Worked example

Find the pH of a 0.5 mol dm^{-3} solution of HNO_3.

HNO_3 is a strong acid so it is completely ionized in water.

$HNO_3 \rightarrow H^+ + NO_3^-$

0.5 0.5

$pH = -\log_{10}[H^+]$

$pH = -\log_{10}0.5$

$pH = 0.30$

Finding the pH

We can rearrange the equation to find the concentration of H^+ ions if we know the pH.

$[H^+] = 10^{-pH}$

Reminder

To get 10^x using a calculator, press the inverse or shift key and then the log key.

Worked example

A strong acid has a pH of 1. What is the value of $[H^+]$?

$[H^+] = 10^{-pH}$

$[H^+] = 10^{-1}$

$[H^+] = 0.1$ mol dm^{-3}

The ionic product of water

Water will naturally partially ionize.

$$H_2O \rightleftharpoons H^+ + OH^-$$

The equilibrium expression for this reaction is

$$K_c = \frac{[H^+]\,[OH^-]}{[H_2O]}$$

As the concentration of water is very large we can consider it to be constant and the expression simplifies to:

$$K_w = [H^+][OH^-]$$

where $K_w = K_c \times [H_2O]$

- K_w is called the **ionic product of water**.
- At 298 K or 25 °C it has a value of 1×10^{-14} $mol^2\ dm^{-6}$.
- The value of K_w changes with temperature.

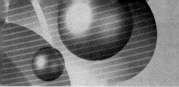

Neutral solutions

In neutral solutions the concentration of H^+ ions is equal to the concentration of OH^- ions

$[H^+] = [OH^-]$
$[H^+]^2 = 1 \times 10^{-14}\ mol^2\ dm^{-6}$
$[H^+] = 1 \times 10^{-7}\ mol\ dm^{-3}$
$pH = -\log_{10}[H^+]$
$pH = -\log_{10}(1 \times 10^{-7})$
$pH = 7$

So neutral solutions have a pH of 7 at 25 °C.

Finding the pH of strong bases

We can use the ionic product of water to find the pH of solutions of strong bases such as sodium hydroxide and potassium hydroxide.

Worked example

Find the pH of 0.2 mol dm⁻³ sodium hydroxide solution. Sodium hydroxide is a strong alkali so it is fully ionized in water.

$NaOH \rightarrow Na^+ + OH^-$

0.2

$K_w = [H^+][OH^-]$

$1 \times 10^{-14}\ mol^2\ dm^{-6} = [H^+] \times 0.2\ mol\ dm^{-3}$

$$[H^+] = \frac{1 \times 10^{-14}\ mol^2\ dm^{-6}}{0{\cdot}2\ mol\ dm^{-3}}$$

$[H^+] = 5 \times 10^{-14}\ mol\ dm^{-3}$

$pH = -\log_{10}[H^+]$

$pH = -\log_{10}(5 \times 10^{-14})$

$pH = 13.3$

Reminder

A high value for $[H^+]$ means a low value for pH.

A low value for $[H^+]$ means a high value for pH.

K_a for weak acids

Organic acids such as methanoic acid, ethanoic acid, and propanoic acid are all weak acids (see section 4.10 on carboxylic acids).

- Weak acids have a pH of around 3.
- Weak acids are only partially ionized in water.
- A weak acid can be represented by HA.

$$HA \rightleftharpoons H^+ + A^-$$

As a result we cannot work out $[H^+]$ directly from the concentration of the weak acid, as we could with strong acids.

- Instead we use the **acid dissociation constant, K_a**

$$K_a = \frac{[H^+][A^-]}{[HA]}$$

- Note that the expression does not include water, H_2O.

When using this expression we assume that:

- $[HA]$ = the initial concentration of the weak acid (even though some of the acid will actually have dissociated).
- $[A^-] = [H^+]$ (even though some of the H^+ ions will actually have come from the ionization of water)

As a result the expression for K_a simplifies to:

$$K_a = \frac{[H^+]^2}{[HA]} \quad \text{or} \quad [H^+]^2 = K_a \times [HA]$$

- This allows us to find the value of $[H^+]$ and from this we can calculate the pH of the solution.

Worked example

Calculate the pH of a 0.2 mol dm⁻³ solution of a weak acid, given that $K_a = 1.7 \times 10^{-5}$ mol dm⁻³.

$$K_a = \frac{[H^+][A^-]}{[HA]}$$

$$K_a = \frac{[H^+]^2}{[HA]}$$

$[H^+]^2 = K_a \times [HA]$

$[H^+]^2 = 1.7 \times 10^{-5}\ mol\ dm^{-3} \times 0.2\ mol\ dm^{-3}$

$[H^+]^2 = 3.4 \times 10^{-6}\ mol^2\ dm^{-6}$

$[H^+] = 1.84 \times 10^{-3}\ mol\ dm^{-3}$

$pH = -\log_{10}[H^+]$

$pH = -\log_{10}(1.84 \times 10^{-3})$

$pH = 2.73$

pK_a

$pK_a = -\log_{10}K_a$
$K_a = 10^{-pK_a}$

- A high value for K_a means a low value for pK_a.
- A low value for K_a means a high value for pK_a.

Weak bases

Ammonium hydroxide, NH_4OH, is a weak base.
It is only partially ionized in water.
$NH_4OH \rightleftharpoons NH_4^+ + OH^-$

Questions

1 Give the definition for pH.
2 What is the pH of the following solutions:
  a 0.1 mol dm⁻³ of HCl
  b 0.5 mol dm⁻³ of HNO_3
  c 0.6 mol dm⁻³ of HCl
3 Name three weak acids.

4.06 Acids and bases 2

OBJECTIVES

By the end of this section you should:

- *remember how to carry out a simple titration*
- *know the typical shapes of pH curves for acid–base titrations between weak and strong acids and bases*
- *know how to select an appropriate indicator*
- *know how buffers work*
- *know how to calculate the pH of acidic buffer solutions*

Titrations

In titrations:

- A known volume of an acid is placed in a flask.
- An **indicator** is added.
- An alkali (a soluble base) is added from a burette.
- The indicator is used to show the endpoint. This is when exactly the right amount of alkali has been added so that neither the acid nor the base is in excess.
  As a result the volume of alkali that had to be added can then be accurately measured.

Titrations can also be carried out by placing the alkali in the flask and adding the acid from the burette.

Indicators

Indicators (HInd) are weak acids. When they are placed in water they partially dissociate:

$$\text{HInd} \rightleftharpoons \text{H}^+ + \text{Ind}^-$$

- HInd is one colour and Ind^- is a different colour.
- If the indicator is placed in an acid the position of equilibrium is driven to the left and the colour of HInd is seen.
- If the indictor is placed in a base the position of equilibrium moves to the right and the colour of Ind^- is seen.
- At the end point of the titration $[\text{HInd}] = [\text{Ind}^-]$
- For methyl orange $\text{p}K_{\text{Ind}} = 3.7$. It is red in acid and yellow in bases.
- For phenolphthalein $\text{p}K_{\text{Ind}} = 9.3$. It is colourless in acid and pink in bases.

Titration curves

Titration curves can be used to show how the pH changes as alkali is added to an acid. The shape of the curves depends on whether the acid and alkali used are weak or strong.

Strong acid and strong base

- A strong acid will have a pH of around 1 and a strong base will have a pH of around 13.
  As a result the graph starts at 1 and ends at 13.
- The graph is vertical between pH 3 and pH 11
- The equivalence point (when neither acid nor alkali is present in excess) is at pH 7, which is the middle of the vertical part of the graph.

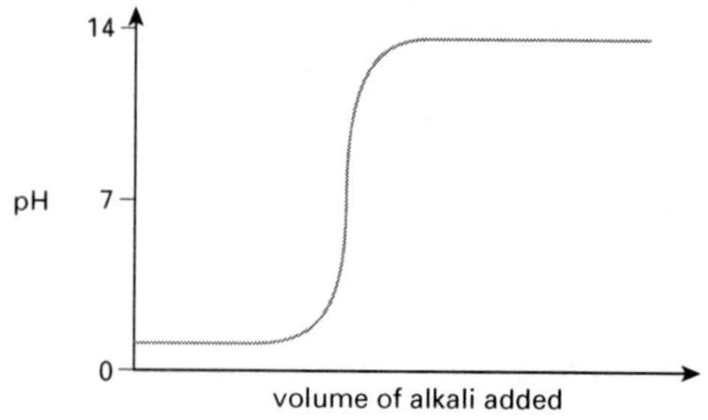

Weak acid and strong base

- A weak acid will have a pH of around 3 and a strong base will have a pH of around 13.
  As a result the graph starts at 3 and ends at 13.
- The graph is vertical between pH 5 and pH 11
- The equivalence point (when neither acid nor alkali is present in excess) is at pH 9.

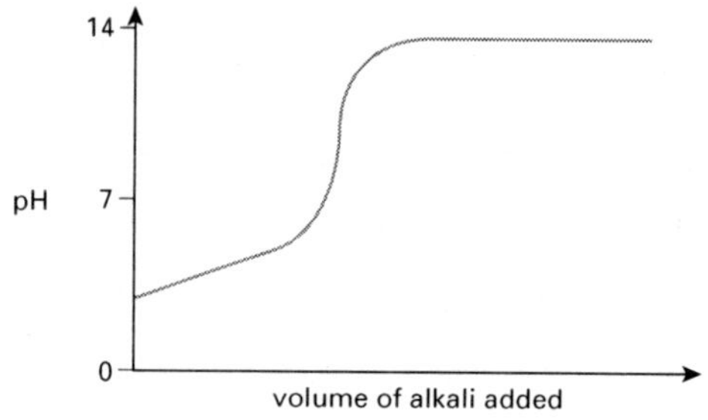

Strong acid and weak base

- A strong acid will have a pH of around 1 and a weak base will have a pH of around 11.
  As a result the graph starts at 1 and ends at 11.
- The graph is vertical between pH 3 and pH 9
- The equivalence point (when neither acid nor alkali is present in excess) is at pH 5.

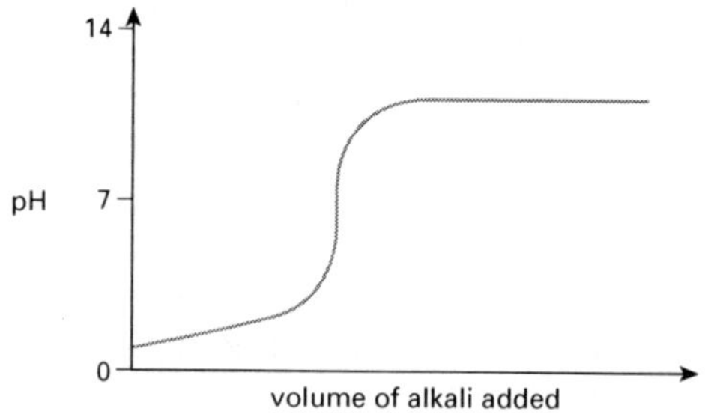

Weak acid and weak base

- A weak acid will have a pH of around 3 and a weak base will have a pH of around 11.
  As a result the graph starts at 3 and ends at 11.

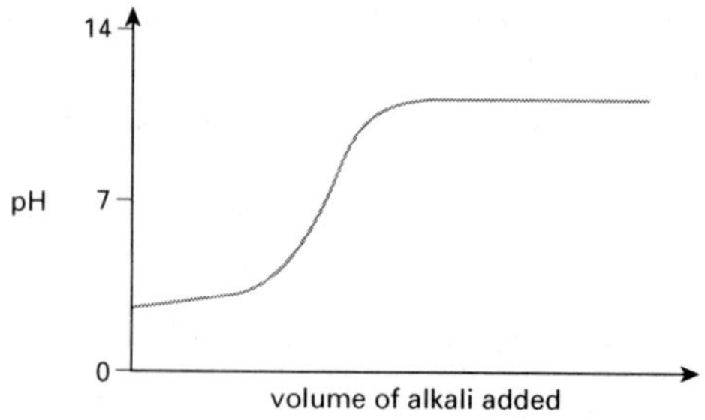

- Unfortunately there is no vertical section for this titration curve, so this method is not suitable for weak acids and weak bases.

Note that titration curves can also be drawn for titrations where acids are added to bases.

Picking the right indicator

Different indictors are available. The type of indictor selected depends on the type of acid and the type of base involved in the titration. The indicator must have a pK_{Ind} value which lies on the vertical part of the graph (see pK_{Ind} values in the indicators section opposite).

- For a strong acid and a strong base either methyl orange or phenolphthalein indicator would be suitable.
- For a weak acid and a strong base phenolphthalein indicator would be suitable.
- For a strong acid and a weak base methyl orange indicator would be suitable.

Buffer solutions

A **buffer solution** resists a change in pH when small amounts of acid or base are added to it.

Buffers are made of an acid–base conjugate pair.

An **acidic buffer** is made by dissolving the salt of a weak acid in the acid itself, for example sodium ethanoate in ethanoic acid.

This gives the weak acid CH_3COOH and its conjugate base CH_3COO^-.

Acidic buffers have a pH of less than 7. They work best when the concentration of the acid and the conjugate base are similar, for example $[CH_3COOH] = [CH_3COO^-]$

A basic buffer is made by dissolving the salt of a weak base in the base itself, for example ammonium chloride in ammonia solution.

This gives the weak base NH_3 and its conjugate acid NH_4^+.

Basic buffers have a pH of greater than 7.

They work best when the concentration of the base and the conjugate acid are similar, for example $[NH_3] = [NH_4^+]$.

How do acidic buffers work?

As discussed above, an acidic buffer can be made by dissolving sodium ethanoate in ethanoic acid.

When the salt of the weak acid sodium ethanoate is dissolved in the weak acid it is fully ionized.

$$CH_3COONa \rightarrow CH_3COO^- + Na^+$$

This means there is a large reservoir of CH_3COO^- ions.

Because ethanoic acid is a weak acid it is only partially ionized.

In addition, the high concentration of ethanoate ions from the dissolved salt further suppresses the ionization of the acid so very little ethanoic acid will dissociate:

$$CH_3COOH \rightleftharpoons CH_3COO^- + H^+$$

This means there is a large reservoir of CH_3COOH molecules.

- $[CH_3COO^-]$ is given by the concentration of the salt of the weak acid.
- $[CH_3COOH]$ is given by the initial concentration of the weak acid.

Buffer solutions resist changes to pH when small amounts of acid are added because the H^+ ions react with ethanoate ions, CH_3COO^-, to form ethanoic acid molecules:

$$H^+ + CH_3COO^- \rightleftharpoons CH_3COOH$$

Buffer solutions resist changes to pH when small amounts of base are added because the OH^- ions react with ethanoic acid molecules, CH^3COOH to form ethanoate ions and water molecules:

$$CH_3COOH + OH^- \rightleftharpoons CH_3COO^- + H_2O$$

Calculating the pH of an acidic buffer solution

We can work out the concentration of the hydrogen ions and use this to work out the pH of an acidic buffer solution.

$$K_a = \frac{[H^+][A^-]}{[HA]}$$

- [HA] is the concentration of weak acid. We assume that it is the same as the initial concentration of the acid.

$$[H^+] = \frac{K_a \times [HA]}{[A^-]}$$

- $[A^-]$ is the concentration of conjugate base. We assume that it is the same as the concentration of the salt.

$$pH = -\log_{10}[H^+]$$

Worked example

Calculate the pH of a buffer made by mixing 0.30 moles of a weak acid, HA, with 0.2 moles of its salt, NaA in 2.0 dm³ of solution.

$K_a = 4.0 \times 10^{-6}$ *mol dm⁻³*

The salt is fully ionized $NaA \rightarrow Na^+ + A^-$

The weak acid is partially ionized $HA \rightleftharpoons H^+ + A^-$

$$K_a = \frac{[H^+][A^-]}{[HA]}$$

$$\text{Concentration of } A^- = \frac{0{\cdot}20\ \text{mol}}{2{\cdot}0\ \text{dm}^3} = 0{\cdot}1\ \text{mol dm}^{-3}$$

$$\text{Concentration of HA} = \frac{0{\cdot}30\ \text{mol}}{2{\cdot}0\ \text{dm}^3} = 0{\cdot}15\ \text{mol dm}^{-3}$$

$$4 \times 10^{-6}\ \text{mol dm}^{-3} = \frac{[H^+] \times 0{\cdot}1\ \cancel{\text{mol dm}^{-3}}}{0{\cdot}15\ \cancel{\text{mol dm}^{-3}}}$$

$$[H^+] = \frac{4 \times 10^{-6}\ \text{mol dm}^{-3} \times 0{\cdot}15}{0{\cdot}1} = 6 \times 10^{-6}\ \text{mol dm}^{-3}$$

$$pH = -\log_{10}[H^+] = 5{\cdot}2$$

Useful buffers

Non-toxic buffers are used to maintain suitable pH values in a wide range of substances including: shampoos; baby lotions; eye drops; washing powders; skin creams.

Blood is also buffered by hydrogencarbonate ions and also by haemoglobin and plasma.

Questions

1 Suggest the name of an indicator that would be suitable for a titration between a strong acid and a weak base.

2 What is a buffer solution?

3 Suggest how you could make an acidic buffer solution.

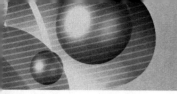

4.07 Naming organic compounds

OBJECTIVES

By the end of this section you should know:

- *the terms molecular formula, structural formula, displayed formula, homologous series, functional groups*
- *how to name alkanes, alkenes, and haloalkanes*
- *about positional isomers*
- *how to name the organic compounds covered by the A2 syllabus*

- Members of a **homologous series** have the same general formula. For example, alkenes have the general formula C_nH_{2n}. Each member of a homologous series differs from the previous member by the addition of a CH_2 group.
- A **functional group** is an atom, or group of atoms, which determines how the compound will react. Members of the same homologous series have the same functional group.
- The **molecular formula** shows the type and number of each type of atom in a compound. Propane has the molecular formula C_3H_8.
- The **structural formula** of a compound shows the number of each type of atom. It also shows how the atoms are joined together.
- Propane has the structural formula $CH_3CH_2CH_3$.
- Pentane has the structural formula $CH_3CH_2CH_2CH_2CH_3$. This can be simplified to the shortened structural formula $CH_3(CH_2)_3CH_3$.
- The **displayed formula** shows the number of each type of atom and the bonds in the compound. The bonds are shown as straight lines.

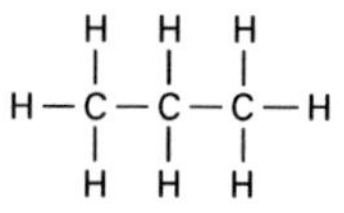
propane

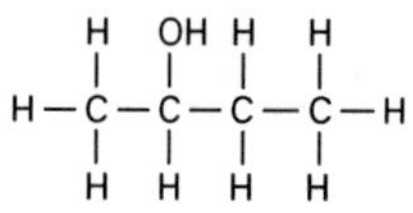
butan-2-ol

The **skeletal formula** is a useful way of representing the shape of a molecule. Single covalent bonds between two carbon atoms are shown as single lines. Note that symbols for carbon atoms and hydrogen atoms bonded to carbon atoms are not shown but the symbols for other atoms are added.

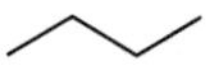
propane

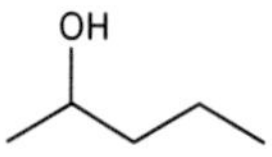
butan-2-ol

Naming unbranched organic compounds

The names of unbranched organic compounds have three parts.

- The first part tells you the number of carbon atoms in the compound.

number of carbon atoms	prefix (first part of name)
1	meth-
2	eth-
3	prop-
4	but-
5	pent-
6	hex-

- The middle part tells you the bonding in the chain
  -an- means there are only single bonds
  -en- means there is a double bond in the carbon chain.
- The last part of the name tells you which functional groups are attached to the carbon chain.
  -ol means that a hydroxyl, –OH, group is attached to the carbon chain.
  -al means there is a carbonyl, >C=O, group at the end of the carbon chain.
  -oic acid means there is a carboxyl, COOH, group atached to the carbon chain.
  -one means there is a carbonyl, >C=O, group in the carbon chain but not at the end.
- We use a number to give the position of the functional group along the carbon chain. When numbers are needed we add hyphens between the parts of the names.

Naming branched organic compounds

- To name an organic compound with a branched chain, first find the number of carbon atoms in the main (longest) chain.
- The branches that come off the main chain are called **side chains**.

number of C atoms in side chain	side chain	name
1	$-CH_3$	methyl
2	$-CH_2CH_3$	ethyl
3	$-CH_2CH_2CH_3$	propyl
4	$-CH_2CH_2CH_2CH_3$	butyl

- We use numbers to give the positions of the side chains.

Questions

1 Look at the diagram below.

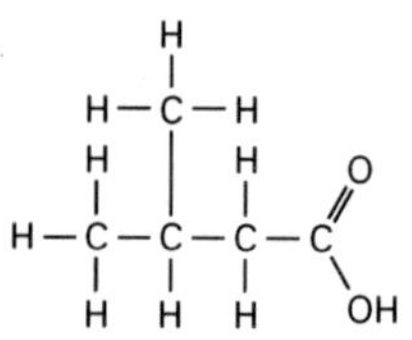

a How many carbon atoms are in the main chain of this molecule?

b Circle the functional group in this compound.

c Name this compound.

			example		functional group
class of compound	prefix	suffix	name	displayed formula	
alkane		ane	propane	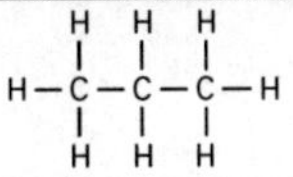	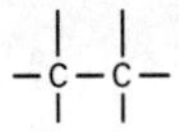
alkene		ene	propene	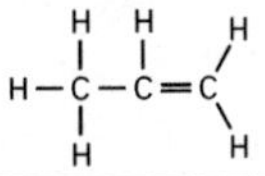	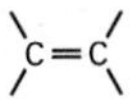
haloalkane	halogeno		1-chloropropane	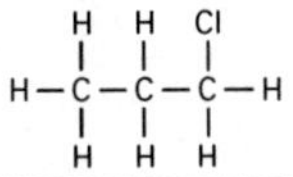	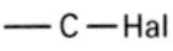
primary alcohol		ol	propan-1-ol	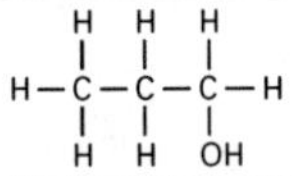	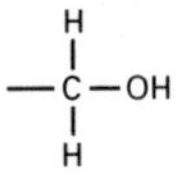
secondary alcohol		ol	propan-2-ol	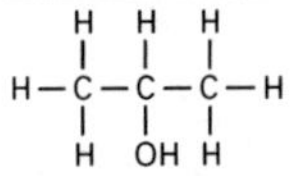	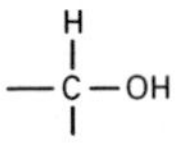
tertiary alcohol		ol	methylpropan-2-ol	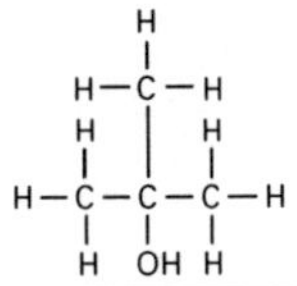	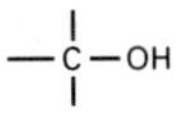
nitrile	cyano	nitrile	propanenitrile	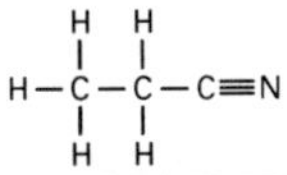	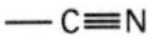
amine	amino	amine	aminopropane	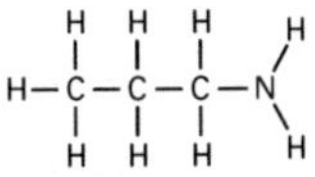	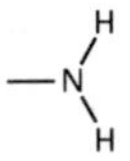
aldehydes		al	propanal	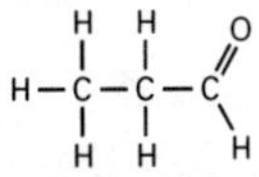	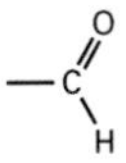
ketones		one	propanone	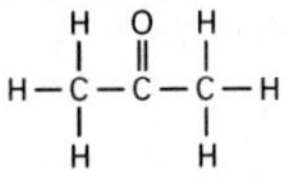	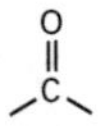
carboxylic acids		oic acid	propanoic acid	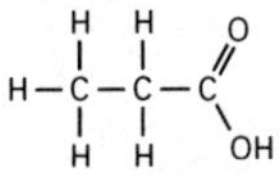	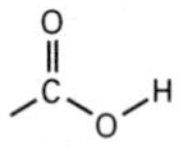
acyl chloride		oyl chloride	propanoyl chloride	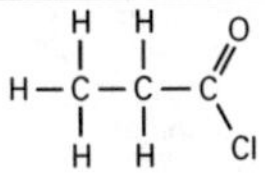	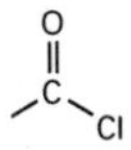
amide		amide	propanamide	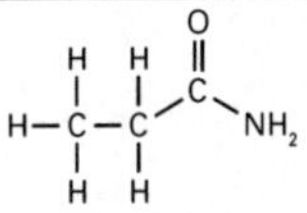	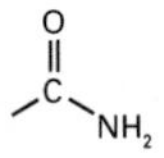
ester		oate	ethylpropanoate	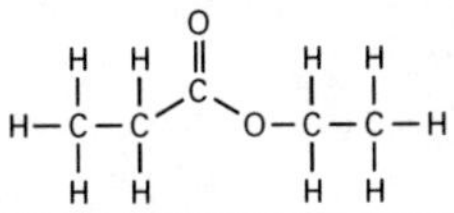	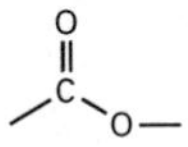
acid anhydride		anhydrite	propanoic anhydride	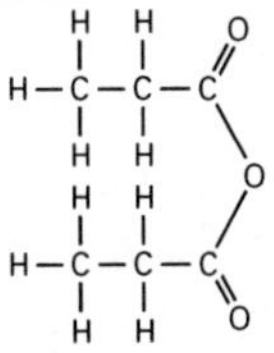	
arenes and aromatics	phenyl	benzene	phenylamine nitrobenzene	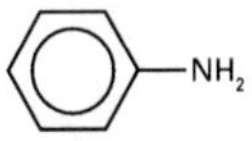 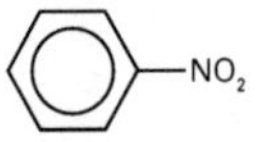	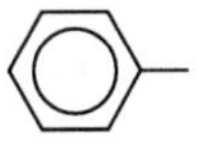

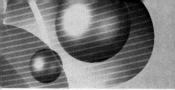

4.08 Isomers

OBJECTIVES

By the end of this section you should know:

- *how to draw chain and position isomers*
- *that alkenes can exhibit E-Z isomerism*
- *what structural isomers are*
- *that E-Z isomerism and optical isomerism are forms of stereoisomerism*
- *the meanings of the terms enantiomers and racemic mixture*
- *that an asymmetric carbon gives rise to optical isomers*
- *that optical isomers have non-superimposable mirror images*
- *that different optical isomers may have very different effects*

Structural isomers

Structural isomers have the same number of each type of atom (same molecular formula) but a different structural arrangement.

Worked example

Three carbonyl compounds have the same molecular formula, C_4H_8O. Name and give the displayed formula of these three compounds.

The compounds contain a >C=O group.

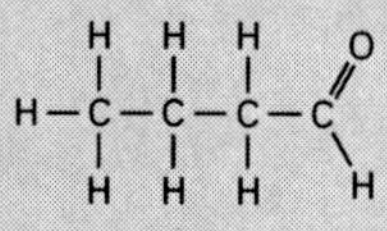
butanal

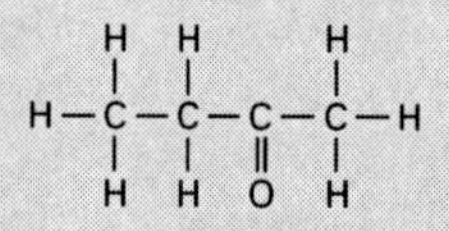
butanone

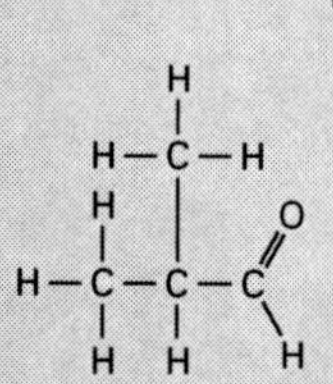
methyl propanal

Chain isomers

Chain isomers are structural isomers in which the carbon atoms are arranged in different ways. For example, methylpropane and butane are chain isomers.

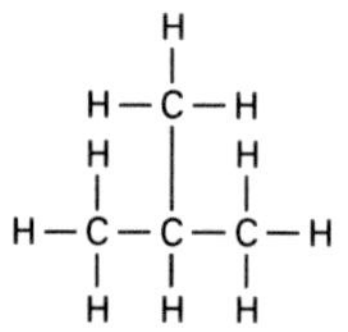
methylpropane

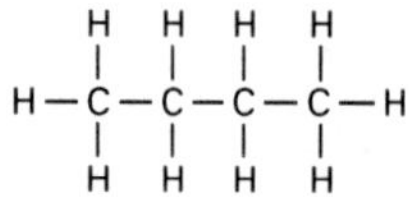
butane

Position isomers

Position isomers are structural isomers in which the same functional group is positioned at different places along the carbon chain.

Butan-1-ol and butan-2-ol are position isomers.

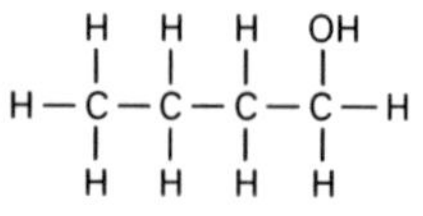
butan-1-ol

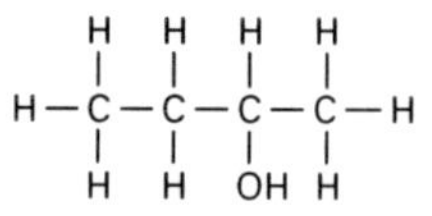
butan-2-ol

Stereoisomers

Stereoisomers have the same molecular formula and the same structural formula but the atoms have a different arrangement in space. Stereoisomers include:

- **Geometric** or *E-Z* isomers (these have a C=C bond)
- **Optical isomers** (these have an **asymmetric** carbon atom)

Geometric or E-Z *Isomers*

- Although there is normally free rotation about a C–C bond there is restricted rotation about a C=C bond. As a result it is possible for alkene molecules to exist as stereoisomers.

For stereoisomers (*E-Z* isomers) to exist a molecule must have:

- a C=C bond
- two different groups attached to the carbon atoms involved in the double bond

Example

But-2-ene has two stereoisomers.

The *E* isomer is

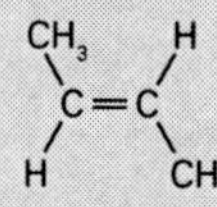

E comes from the German word *entgegen*. It means opposite.

The *Z* isomer is

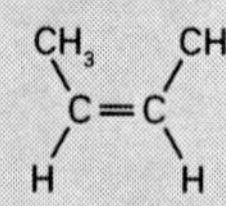

Z comes from the German word *zusammen*. It means together.

Notice that but-1-ene does not exit as stereoisomers.

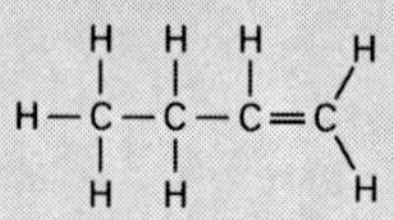

This is because one of the carbon atoms involved in the double bond is attached to two identical groups (two hydrogen atoms).

Deciding the priority of groups

To decide whether to name an isomer *E* or *Z* the groups attached to the carbon atoms in the double bond must be examined. Each group is given a **priority**.

- H is given the lowest priority
- then CH_3
- then C_2H_5 and so on

You then decide whether the groups are attached above or below the double bond.

Example

This is one isomer of 3-methylpent-2-ene

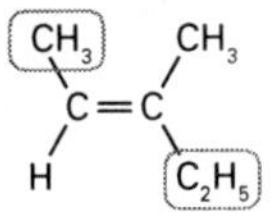

The highest priority groups on each carbon atom are circled.

One of the highest priority groups is above the double bond and the other is below the double bond so this is (*E*)-3-methylpent-2-ene.

Here is the other isomer.

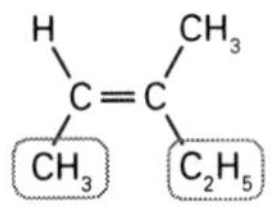

Again the highest priority groups on each carbon atom are circled.

Both of the highest priority group are below the double bond so this is (*Z*)-3-methylpent-2-ene.

Worked example

Name and give the displayed formula of the three non-cyclic structural isomers of C_4H_8.

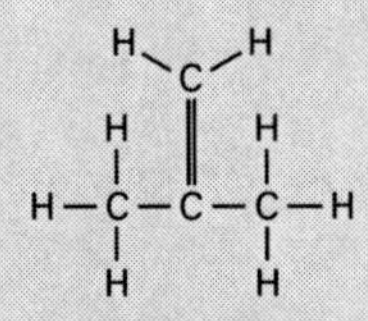

methylpropene

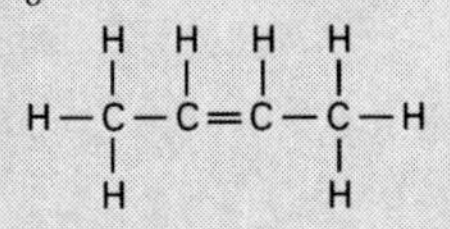

but-2-ene

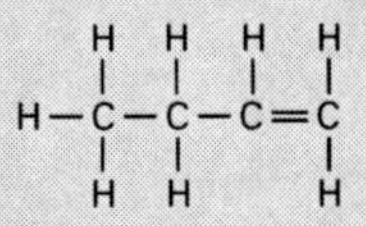

but-1-ene

Haloalkenes

In haloalkenes the order of priority for the groups attached to the carbon atoms in the double bond is iodine>bromine>chlorine>fluorine

Optical isomers

Optical isomers are another type of stereoisomers. Optical isomers are **non-superimposable** mirror images of each other. The different isomers are called **enantiomers**. Optical isomers do not have to contain C=C bonds. They occur when a molecule contains a carbon atom which is attached to four different groups. This carbon atom is called an asymmetric carbon and it is a chiral centre.

Example

This is the displayed formula of 2-iodobutane

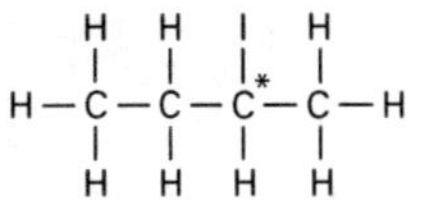

* asymmetric carbon – a carbon atom attached to four different groups.

2-Iodobutane has two enantiomers. Notice how they are non-superimposable mirror images of each other.

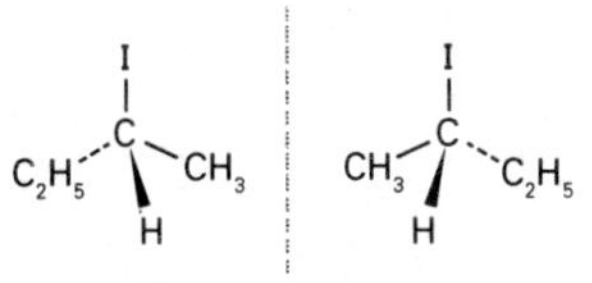

Distinguishing between enantiomers of a compound

Although they are chemically identical we can tell enantiomers apart by their optical activity. One enantiomer will rotate the plane of plane-polarized light clockwise while the other enantiomer will rotate the plane anticlockwise.

Racemic mixtures

A **racemic mixture** contains equal amounts of each enantiomer. This means that racemic mixtures do not show optical activity as the rotating effect of one enantiomer is cancelled out by the other enantiomer.

Synthetic and natural organic molecules (HSW)

Chemicals produced in the laboratory, such as synthetic amino acids, are not optically active because equal amounts of each enantiomer are produced.

However, if the same amino acid is produced by a living organism it will be optically active because only one enantiomer will be produced. This has important consequences in the manufacture of drugs. If a drug containing a chiral centre is produced in the laboratory it is likely that both enantiomers will be produced. Although one of these enantiomers may be very beneficial the other may be harmful. This was the case with the drug thalidomide. Introduced in the late 1950s one enantiomer relieved the symptoms of morning sickness. Unfortunately the other enantiomer caused serious birth defects. Although thalidomide is still used to treat leprosy, it is never given to pregnant women.

Questions

1. Does pent-1-ene have two stereoisomers? Explain your answer.
2. What is the difference between the E and the Z forms of but-2-ene?
3. Will a compound that has optical isomers always be optically active? Explain your answer.

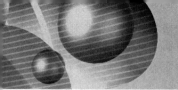

4.09 Aldehydes and ketones

OBJECTIVES

By the end of this section you should know:

- *that primary alcohols can be oxidized to form aldehydes*
- *that secondary alcohols can be oxidized to form ketones*
- *how to use Fehling's solution and Tollens' reagent to distinguish between aldehydes and ketones*
- *that aldehydes are readily oxidized to carboxylic acids*
- *that aldehydes can be reduced to primary alcohols and ketones can be reduced to secondary alcohols*
- *about the mechanism of the nucleophilic addition reactions of aldehydes and ketones*
- *about the hazards of synthesis using HCN or KCN*

Aldehydes

Aldehydes and ketones both contain a **carbonyl**, >C=O, group.

Aldehydes have the general formula

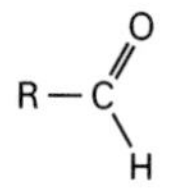

Examples include

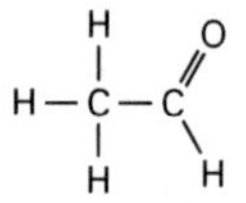

ethanal

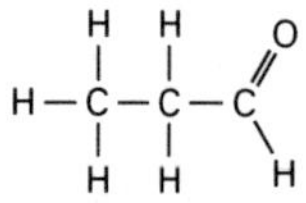

propanal

Ketones

Ketones have the general formula

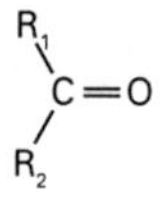

Examples include

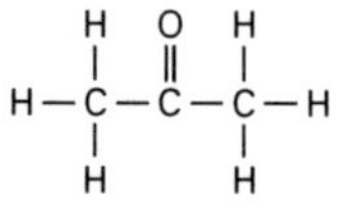

propanone

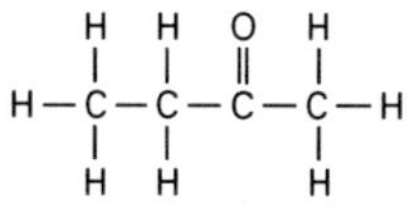

butanone

Ketones that have five or more carbon atoms have position isomers, so numbers are used to indicate the position of the carbonyl group.

Oxidation and reduction

Oxidation can be described as:

- loss of electrons
- gain of oxygen atoms
- loss of hydrogen atoms

Reduction can be described as:

- gain of electrons
- loss of oxygen atoms
- gain of hydrogen atoms

Worked example

Look at the displayed formula for the compound, ***A****, below.*

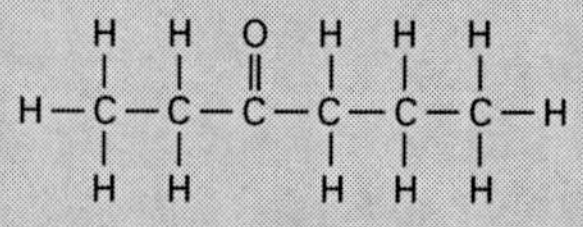

compound A

1 *Identify the functional group present in this compound.*

2 *Name the class of compound that* ***A*** *belongs to.*

3 *Name compound* ***A****.*

1 A contains the carbonyl, >C=O, functional group.

2 A is a ketone

3 hexan-3-one

Aldehydes are made by the oxidation of primary alcohols while ketones are made by the oxidation of secondary alcohols. Aldehydes are readily oxidized, while ketones cannot readily be oxidized.

primary alcohol $\underset{\text{reduction}}{\overset{\text{oxidation}}{\rightleftharpoons}}$ aldehyde $\underset{\text{reduction}}{\overset{\text{oxidation}}{\rightleftharpoons}}$ carboxylic acid

secondary alcohol $\underset{\text{reduction}}{\overset{\text{oxidation}}{\rightleftharpoons}}$ ketone

Distinguishing between aldehydes and ketones

Aldehydes are readily oxidized by oxidizing agents such as **acidified potassium dichromate(VI)**, $K_2Cr_2O_7$, to carboxylic acids.

For example propanal is oxidized to propanoic acid:

$$CH_3CH_2CHO + [O] \rightarrow CH_3CH_2COOH$$

Notice how the symbol [O] is used to represent the oxidizing agent.

During the reaction the dichromate(VI) ions are reduced to chromium(III) ions, so there is a colour change from orange to green.

By contrast, ketones are not readily oxidized.

These observations can be used to distinguish between aldehydes and ketones.

Fehling's solution

Fehling's solution contains a Cu^{2+} complex. It is a blue solution. When Fehling's solution is added to an aldehyde and then gently warmed in a water bath at around 60 °C the aldehyde is oxidized to a carboxylic acid.

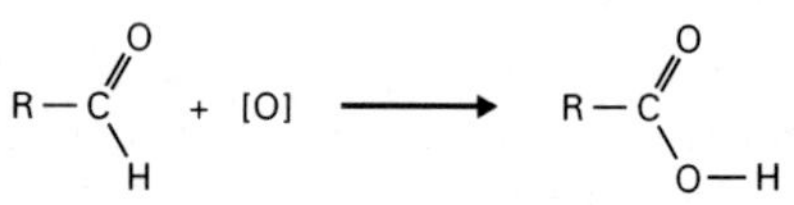

while the Cu^{2+} ions are reduced to a brick-red precipitate of copper(I) oxide, Cu_2O:

$$2Cu^{2+}(aq) + 2e^- + 2OH^-(aq) \rightarrow Cu_2O(s) + H_2O(l)$$

Ketones are not normally oxidized, so if a ketone is warmed with Fehling's solution there is no reaction and so no colour change.

Tollens' reagent

Tollens' reagent contains ammoniacal silver nitrate, $AgNO_3$. When Tollens' reagent is added to an aldehyde and then gently warmed in a water bath at around 60 °C the aldehyde is oxidized to a carboxylic acid, while the silver ions, Ag^+, are reduced to silver atoms, Ag, which form a layer called a silver mirror:

$$Ag^+(aq) + e^- \rightarrow Ag(s)$$

If a ketone is warmed with Tollens' reagent there is no reaction and so no colour change.

Reduction of aldehydes

Aldehydes can be reduced to primary alcohols by reducing agent such as sodium tetrahydridoborate(III) , $NaBH_4$, also known as sodium borohydride.

The reducing agent is added and then heated under reflux. Water or ethanol can be used as solvents.

Lithium tetrahydridoaluminate(III), $LiAlH_4$, also known as lithium aluminium hydride, can also be used. This is a more powerful reducing agent so heating is not required and a dry (without water) ether solvent must be used.

$NaBH_4$ and $LiAlH_4$ reduce the carbonyl, C=O, bond but they do not reduce C=C bonds. They both produce hydride, H^-, ions.

The general equation for this reaction is:

$$RCHO + 2[H] \rightarrow RCH_2OH$$

Notice how the symbol [H] is used to represent the reducing agent.

Reduction of ketones

Ketones can also be reduced by reducing agents such as sodium tetrahydridoborate(III) or lithium tetrahydridoaluminate(III). Ketones are reduced to secondary alcohols.

The general equation for this reaction is:

$$R_1COR_2 + 2[H] \rightarrow R_1CH(OH)R_2$$

Example

The ketone propanone can be reduced to the secondary alcohol propan-2-ol:

$$CH_3COCH_3 + 2[H] \rightarrow CH_3CH(OH)CH_3$$

The carbonyl group

Aldehydes and ketones both contain the carbonyl group.

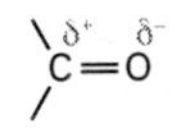

This group is unsaturated and polar and undergoes nucleophilic addition reactions.

Nucleophilic addition reactions

The mechanism for the reduction of aldehydes is shown below.

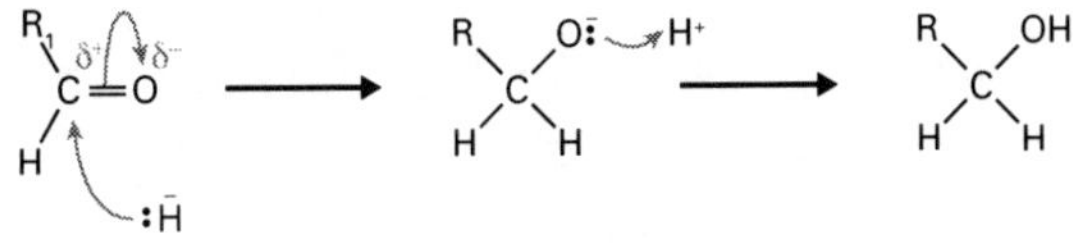

The mechanism for the reduction of ketones is shown below.

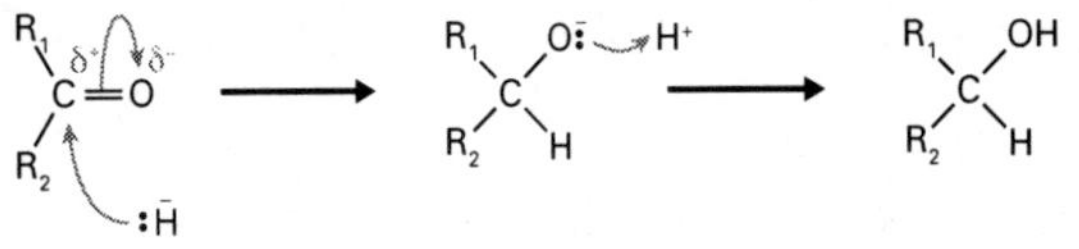

Notice how

- the relevant dipoles and lone pairs of electrons are included
- curly arrows are used to show the movement of pairs of electrons

Hydroxynitriles

The cyanide ion, –CN, is a nucleophile.

Aldehydes and ketones react with hydrogen cyanide to produce hydroxynitriles.

Example

- propanal + hydrogen cyanide → 2-hydroxybutanenitrile

$$CH_3CH_2CHO + HCN \rightarrow CH_3CH_2C(OH)CN$$

This is another example of a nucleophilic addition mechanism. The mechanism for the reaction of aldehydes with hydrogen cyanide is shown below.

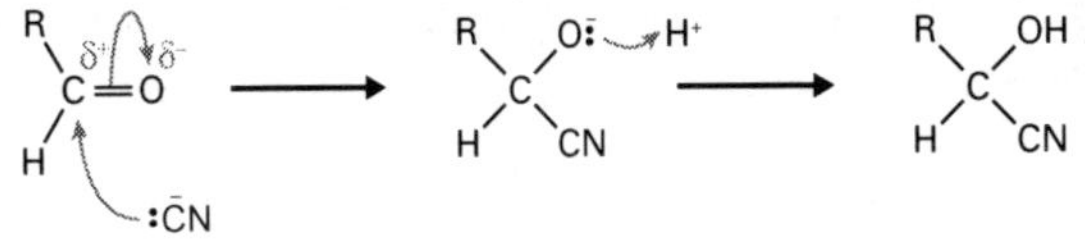

The mechanism for the reaction of ketones with hydrogen cyanide is shown below.

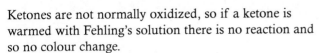

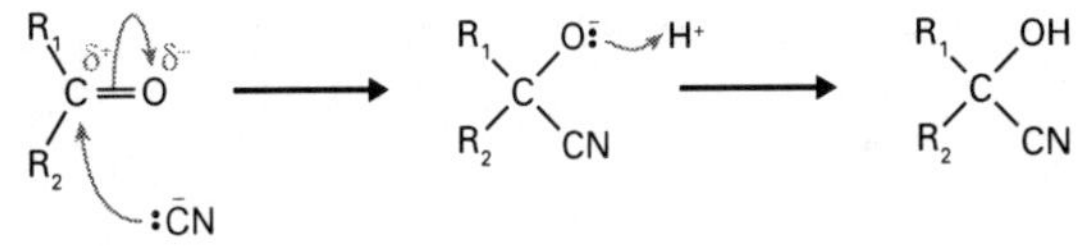

- Notice how the length of the carbon chain has been increased by one.

Hazards of using hydrogen cyanide and potassium cyanide

Hydrogen cyanide and cyanide ions are toxic, so experiments involving these chemicals are not carried out at this level.

Hydrogen cyanide is a covalent molecule, so in a sample of hydrogen cyanide gas there would only be a very low concentration of cyanide ions. Cyanide ions are usually produced from potassium cyanide.

Questions

1 Describe how you could use Tollens' reagent to distinguish between an aldehyde and a ketone.

2 Name the carboxylic acid formed when ethanal is oxidized.

3 Give the equation for the reduction of butanal.

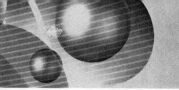

4.10 Carboxylic acids

OBJECTIVES

By the end of this section you should know:

○ *that carboxylic acids are weak acids which react with carbonates to release carbon dioxide*

○ *that carboxylic acids react with alcohols in the presence of a catalyst to form esters*

○ *some uses of esters*

The carboxyl group

Carboxylic acids contain the carboxyl group, –COOH.

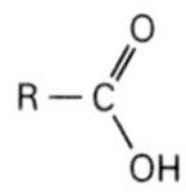

$$R-C\begin{smallmatrix}O\\\\OH\end{smallmatrix}$$

The **carboxyl** group contains both the hydroxyl group, –OH, and the carbonyl group, >C=O.
The presence of both groups modifies the properties of each so that carboxylic acids have their own set of chemical properties.

$$\overset{\delta^+}{\underset{}{C}}=\overset{\delta^-}{O} \qquad -\overset{\delta^+}{C}-\overset{\delta^-}{O}H$$

Both the hydroxyl group and the carbonyl group are polar. This means that the carboxyl group is even more polar than either a carbonyl or an alcohol.

$$-C\begin{smallmatrix}O^{\delta^-}\\\\OH\end{smallmatrix}$$

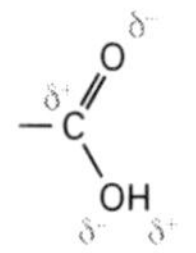

Naming carboxylic acids

Unbranched **carboxylic acids** are named according to the number of carbon atoms they have.

Example

$$H-\overset{H}{\underset{H}{C}}-\overset{H}{\underset{H}{C}}-C\begin{smallmatrix}O\\\\OH\end{smallmatrix}$$

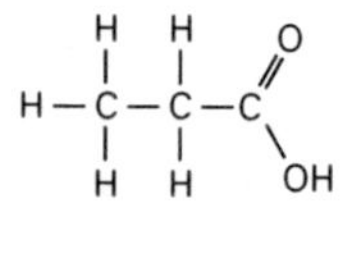

This is propanoic acid.

If the carboxylic acid has a branched chain we use numbers to indicate the position of the side chains. We number the carbon atoms starting from the carbon atom in the carboxyl group.

$$H-\overset{H}{\underset{H}{C}}-\overset{H}{\underset{H}{C}}-\overset{CH_3}{\underset{H}{C}}-C\begin{smallmatrix}O\\\\OH\end{smallmatrix}$$

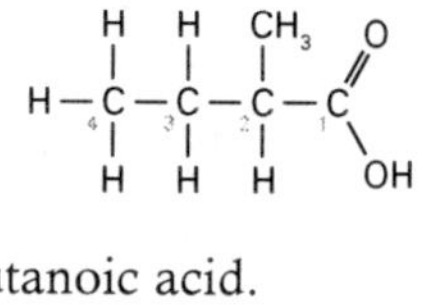

This is 2-methyl butanoic acid.

Making carboxylic acids

Carboxylic acids are made by the oxidation of primary alcohols or aldehydes.
Propan-1-ol is oxidized by potassium dichromate(VI) to propanal:

$$CH_3CH_2CH_2OH + [O] \rightarrow CH_3CH_2CHO + H_2O$$

If the oxidizing agent is in excess, the propanal is further oxidized to propanoic acid:

$$CH_3CH_2CHO + [O] \rightarrow CH_3CH_2COOH$$

Acidic reactions of carboxylic acids

Carboxylic acids are weak acids. They partially dissociate in water.

Example

Ethanoic acid is a weak acid:

$$CH_3COOH \rightleftharpoons CH_3COO^- + H^+$$

Notice that the symbol $\rightleftharpoons$ is used to show that the reaction does not go to completion. In fact, the position of equilibrium lies well to the left.
Carboxylic acids react with carbonates to form a salt, water, and carbon dioxide. This reaction can be used to test for the presence of the carboxyl group.

Example

Ethanoic acid + sodium carbonate → sodium ethanoate + water and carbon dioxide:

$$2CH_3COOH + Na_2CO_3 \rightarrow 2CH_3COONa + H_2O + CO_2$$

This is a neutralization reaction.

Fizzing is seen as the sodium carbonate is added to the carboxylic acid.

Carboxylate ions

Carboxylic acids react to form **carboxylate ions**.

Example

Ethanoic acid reacts with an alkali to form an ethanoate ion.

$$H-\overset{H}{\underset{H}{C}}-C\begin{smallmatrix}O\\\\OH\end{smallmatrix} + OH^- \longrightarrow H-\overset{H}{\underset{H}{C}}-C\begin{smallmatrix}O\\\\O^-\end{smallmatrix} + H_2O$$

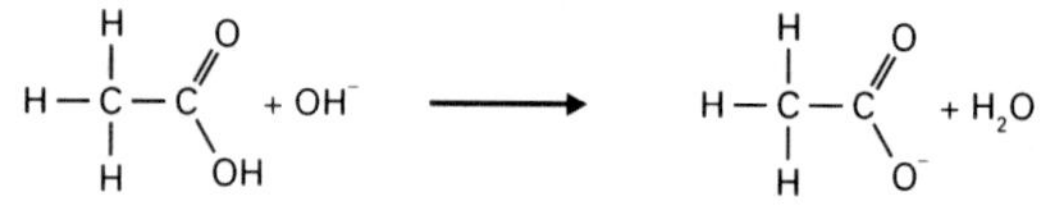

The lone pair of electrons on the oxygen atom and the pi electrons in the C=O bond form a **delocalized system**. This means the negative charge on the oxygen is delocalized and each carbon–oxygen bond is the same length and strength. This stabilizes the carboxylate ion and explains why carboxylic acids are acidic while alcohols, which cannot form delocalized structures, are neutral.

$$\left(H-\overset{H}{\underset{H}{C}}-C\begin{smallmatrix}O\\\\O\end{smallmatrix}\right)^-$$

Esters

Esters contain the group –COOR.

$$R_1-C\begin{smallmatrix}O\\\\O-R_2\end{smallmatrix}$$

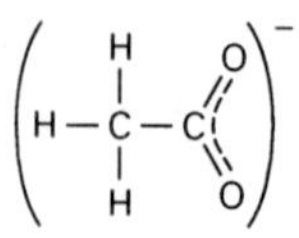

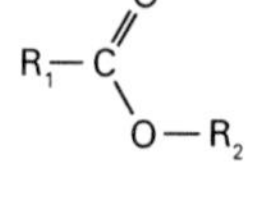

Naming esters

The names of esters are based on the carboxylic acid they are made from and the alkyl group of the alcohol that has replaced the acid proton.

Example

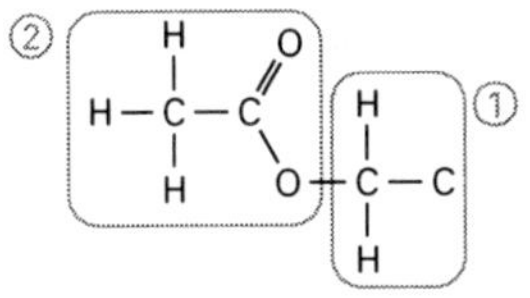

The first part of the name is the alkyl group that has replaced the acid proton; in this case it is methyl.

The second part of the name comes from the carboxylic acid that has been used to make the ester; in this case ethanoic acid.

So this is methyl ethanoate.

Worked example

Name the ester made from propanoic acid and methanol.

Methyl propanoate

Making esters

Esters are made by **esterification** reactions in which carboxylic acids, R_1COOH, react with alcohols, R_2OH, in the presence of a concentrated sulfuric acid or hydrochloric acid catalyst.

$$R_1COOH + R_2OH \overset{H^+}{\rightleftharpoons} R_1COOR_2 + H_2O$$

Note that this reaction is reversible.

In the reaction, a molecule of water is made for each molecule of ester.

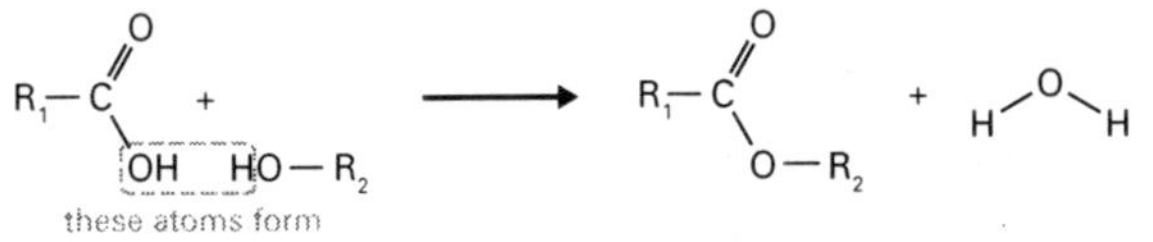

The hydrolysis of esters

The esterification reaction is reversible, and the backwards reaction is called **hydrolysis**.

- Hydrolysis is the breaking down of a compound using water.

$$\text{carboxylic acid} + \text{alcohol} \underset{\text{hydrolysis}}{\overset{\text{esterification}}{\rightleftharpoons}} \text{ester} + \text{water}$$

The reaction can be catalysed by a solution of a dilute acid or of a dilute alkali.

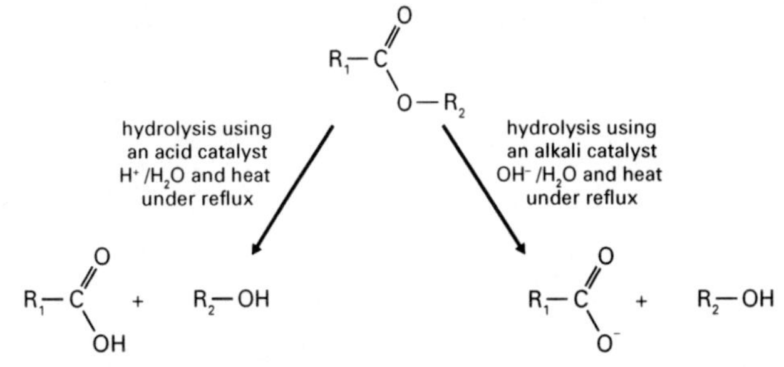

Hydrolysis using an alkali is the preferred method because the reaction happens more quickly. In this method, as the carboxylic acid is made it reacts with the alkali catalysts to form the salt of the carboxylic acid. As the alkali is being used up it is not strictly speaking a catalyst.

Useful esters

Esters are found in nature as oils and fats.

They are very useful compounds which have pleasant, fruity smells and are often used as synthetic food flavourings. It is often cheaper to produce these flavourings than to extract natural flavourings. However, some people can tell the difference between the synthetic and the natural versions.

Esters are also used as solvents in perfumes and also as plasticizers (chemicals that are added to plastics to make them softer and more flexible).

Vegetable oils and animal fats

Vegetable oils and animal fats are the triesters of

- long-chain carboxylic acids (fatty acids)
- propane-1,2,3-triol (glycerol)

In vegetable oils the alkyl chains are normally unsaturated, while in animal fats the chains are normally saturated.

Fats and oils can be hydrolysed by heating with a solution of aqueous sodium hydroxide or potassium hydroxide.

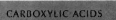

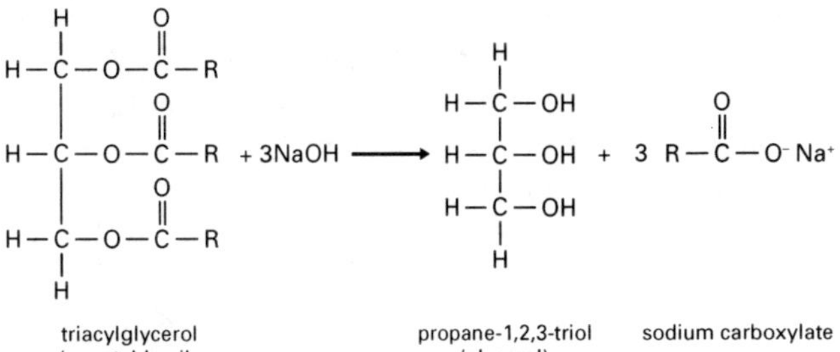

The hydrolysis of vegetable oils and fats produces:

- glycerol – a sweet tasting, colourless liquid
- the sodium or potassium salts of long-chain carboxylic acids.

These salts are used in soaps.

Biodiesel

Biodiesel is a mixture of the methyl esters of long-chain carboxylic acids. It is usually made from vegetable oils.

Questions

1 Name the functional group present in carboxylic acids.

2 Why are esters added to plastics?

3 Name the ester made from ethanoic acid and methanol.

4.11 Acylation 1

OBJECTIVES

By the end of this section you should know:

- *that haloalkanes contain polar bonds*
- *that haloalkanes are susceptible to nucleophilic attack*
- *how to name acyl chlorides*
- *why acyl chlorides are more reactive than haloalkanes*
- *the products of the reactions of acyl chlorides with water, alcohols, ammonia, and primary amines*
- *the mechanisms for the nucleophilic addition–elimination reactions of acyl chlorides*

The acyl group

Acyl groups have the general formula RC=O.

Acylation occurs when the acyl, RC=O, group is added to a compound.

Acyl chlorides

Acyl chlorides contain the functional group –COCl

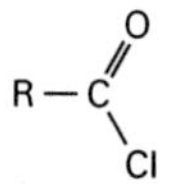

They are reactive organic compounds and are useful in organic synthesis. Compared with carboxylic acids the reactions involving acyl chlorides

- produce better yields of products
- take place more quickly
- take place at room temperature

Acyl chlorides do not occur naturally because they react violently with water.

Naming acyl chlorides

The names of unbranched acyl chlorides are based on the number of carbon atoms they contain (including the carbon atom in the functional group).

The name ends in -oyl chloride.

Examples

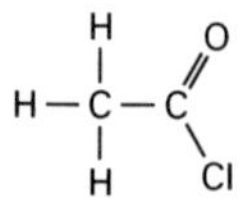

ethanoyl chloride

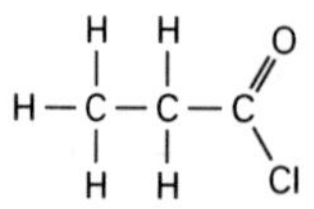

propanoyl chloride

If the acyl chloride has a branched chain we use numbers to indicate the position of the side chains. We number the carbon atoms starting from the carbon atom in the acyl chloride group.

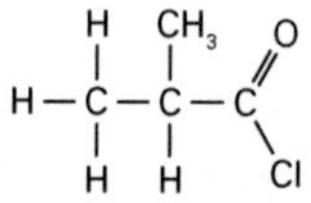

This is 2-methylpropanoyl chloride.

The acyl chloride group

- The oxygen in the carbonyl group and the chlorine in the C–Cl group are both strongly electron-withdrawing.
- This leads to a relatively large δ^+ charge on the carbonyl carbon.

  As a result acyl chlorides are more reactive than chloroalkanes.

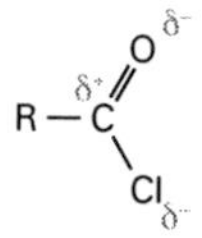

Because of the large δ^+ charge on the carbonyl carbon, acyl chlorides are more susceptible to attack from **nucleophiles** than haloalkanes.

Reminder

Nucleophiles are species that can donate a lone pair of electrons to form a new covalent bond.

Making acyl chlorides

Because of their very high reactivity, acyl chlorides are normally made in situ (i.e. in the place where they are needed). The desired nucleophile is then added. They can be made by reacting phosphorus(V) chloride (also known as phosphorus pentachloride), PCl_5, with a carboxylic acid at room temperature in a nucleophilic substitution reaction.

For example, ethanoyl chloride is made in the laboratory by this reaction:

$$CH_3COOH + PCl_5 \rightarrow CH_3COCl + POCl_3 + HCl$$

Note that the reaction is vigorous, and misty white fumes of hydrogen chloride are produced.

Reactions of acyl chlorides

Acyl chlorides react with:

- water to produce **carboxylic acids**
- alcohols to produce **esters**
- ammonia to produce **primary amides**
- primary amines to produce **N-substituted amides** (or secondary amides)

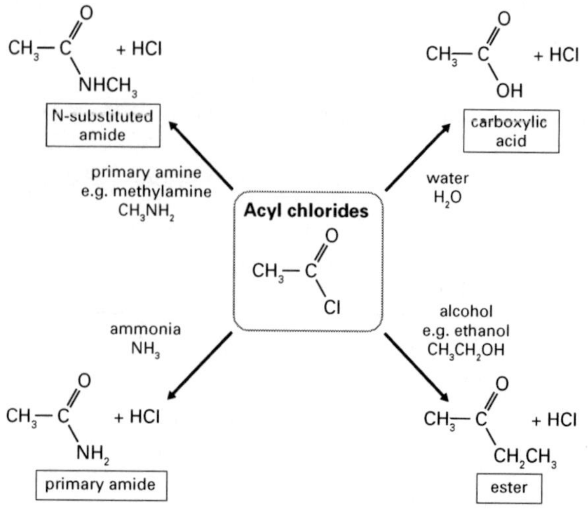

Worked example

1 Give the displayed formula and name of the organic product formed by each of the following reactions.

a *ethanoyl chloride with ammonia*

b *propanoyl chloride with water*

c *ethanoyl chloride with methanol*

d *ethanoyl chloride with methylamine*

a

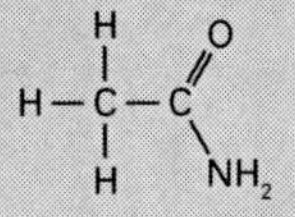

ethanamide

b

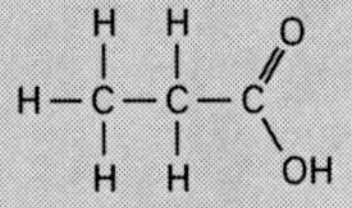

propanoic acid

c

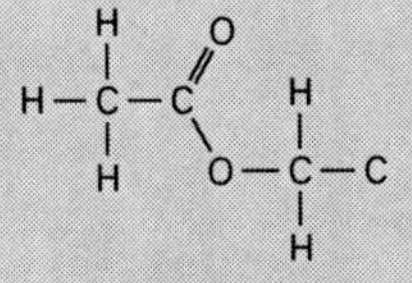

methyl ethanoate

d

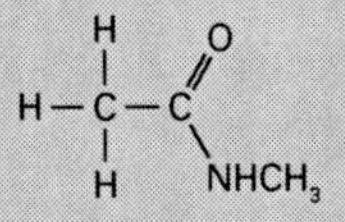

N-methylethanamide

Problems with using acyl chlorides

Although acyl chlorides are useful, chemists sometimes prefer to use acid anhydrides in synthesis.

- Acyl chlorides are more expensive than acid anhydrides.
- Acyl chlorides are more easily hydrolysed.
- When acyl chlorides react they produce fumes of hydrogen chloride.

Nucleophilic addition–elimination reactions

Acyl chlorides react with nucleophiles by an **addition–elimination mechanism.**

Step 1 The nucleophile attacks the δ^+ carbon atom in the –COCl functional group.

Step 2 The chloride ion and a hydrogen atom are removed.

Overall The chlorine atom in the acyl chloride group is substituted for a different group.

The mechanism for the addition–elimination reaction between an acyl chloride and water.

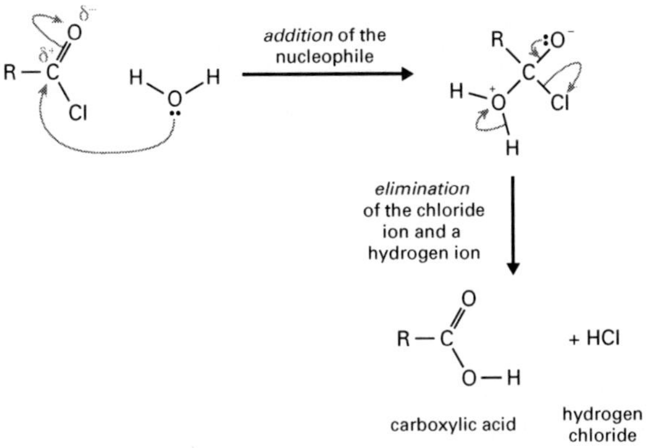

The mechanism for the addition–elimination reaction between an acyl chloride and an alcohol.

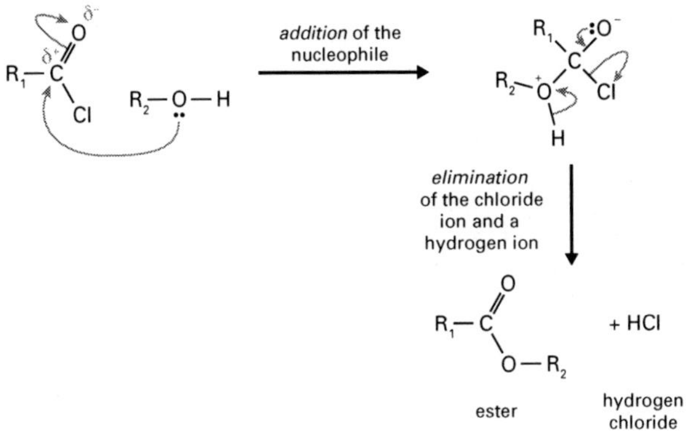

The mechanism for the addition–elimination reaction between an acyl chloride and ammonia.

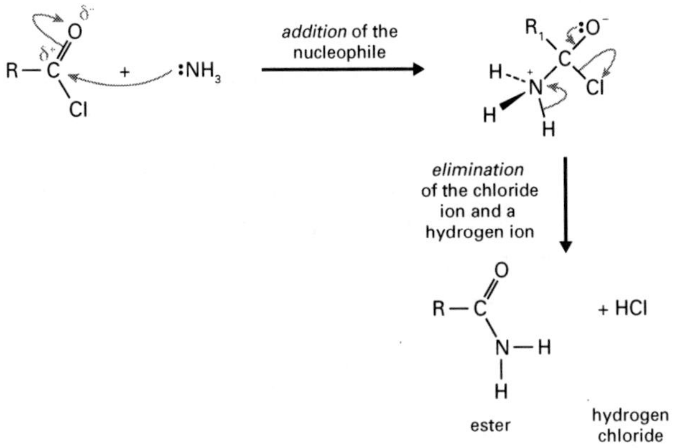

The mechanism for the addition–elimination reaction between an acyl chloride and a primary amine.

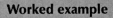

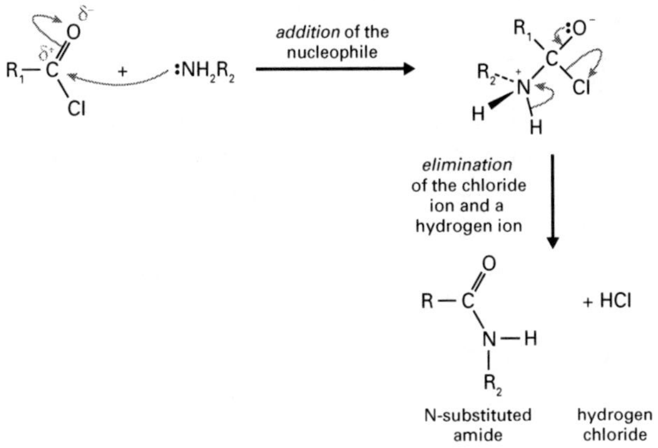

Notice how the relevant dipoles and lone pairs are marked.

Questions

1. Give three advantages of using an acyl chloride rather than a carboxylic acid during organic synthesis.
2. Define a nucleophile.
3. Identify both of the products made when an acyl chloride reacts with an alcohol.

4.12 Acylation 2

OBJECTIVES

By the end of this section you should know:

- *that biodiesel is a mixture of methyl esters of long-chain carboxylic acids*
- *how to name acid anhydrides*
- *the reactions of acid anhydrides with water, alcohols, ammonia, and primary amines*
- *the advantages of using ethanoic anhydride over ethanoyl chloride in the industrial production of aspirin*

Biodiesel

Biodiesel is a fuel used in diesel engines that is not made from crude oil. As well as being a good fuel, biodiesel has good lubricating properties so it helps to reduce engine damage.

Biodiesel is a mixture of the methyl esters of long-chain carboxylic acids. It is usually made from vegetable oils. The vegetable oils are heated with the alcohol methanol and sodium hydroxide.

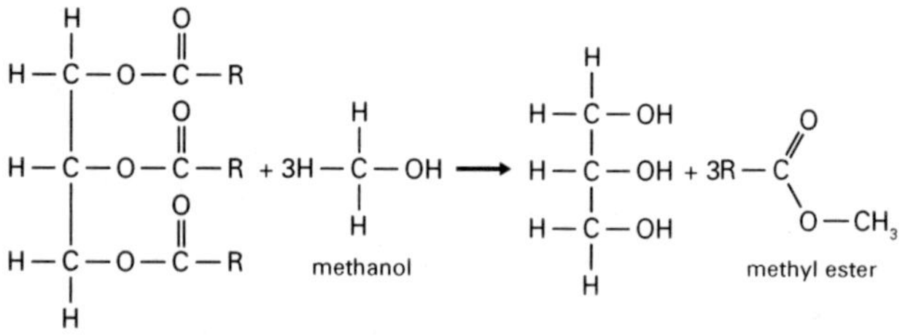

triacylglycerol propane-1,2,3-triol

The biodiesel (consisting of the methyl esters) is less dense and forms a layer on top of the propane-1,2,3-triol. This is called a base-catalysed transesterification reaction.

Rape methyl ester (RME) is made by reacting rape seed oil with methanol. This renewable fuel is almost identical to the non-renewable diesel made from crude oil.

Acid anhydrides

Acid anhydrides contain the functional group –COOCO

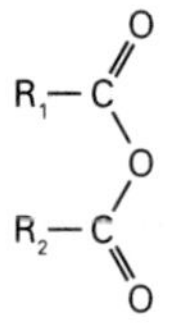

Acid anhydrides are fairly reactive organic compounds and are useful in organic synthesis. Compared with acyl chlorides, acid anhydrides

- are cheaper
- are less susceptible to **hydrolysis**
- react less violently
- do not produce corrosive fumes of hydrogen chloride

As a result acid anhydrides are often preferred to acyl chlorides during organic synthesis.

Naming acid anhydrides

The names of acid anhydrides are based on the names of the carboxylic acids they are derived from.

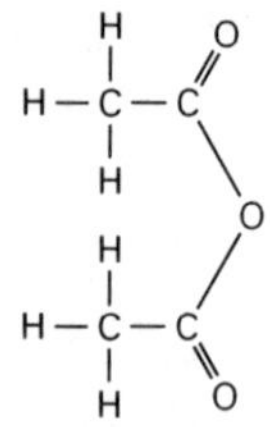

ethanoic anhydride

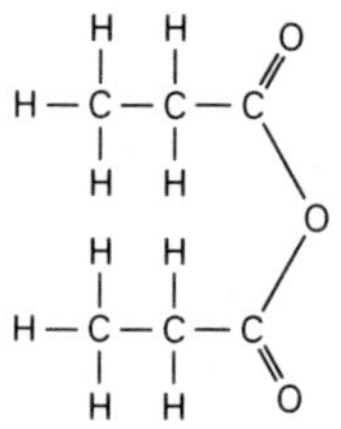

propanoic anhydride

Making acid anhydrides

Symmetric acid anhydrides can be made by **dehydrating** a carboxylic acid.

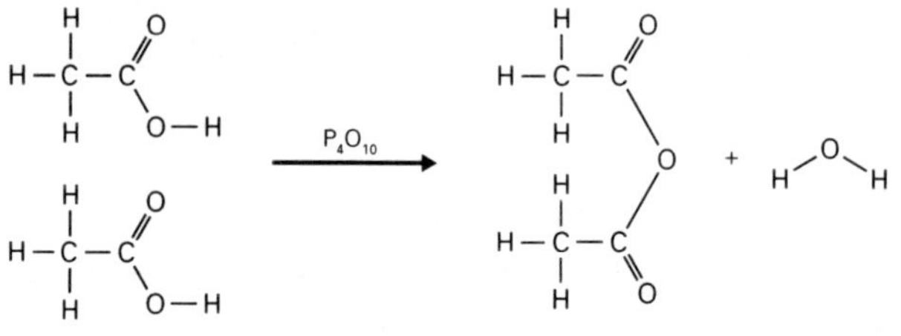

ethanoic anhydride + water

Asymmetric acid anhydrides are made by heating under reflux an acyl chloride with the sodium salt of a carboxylic acid.

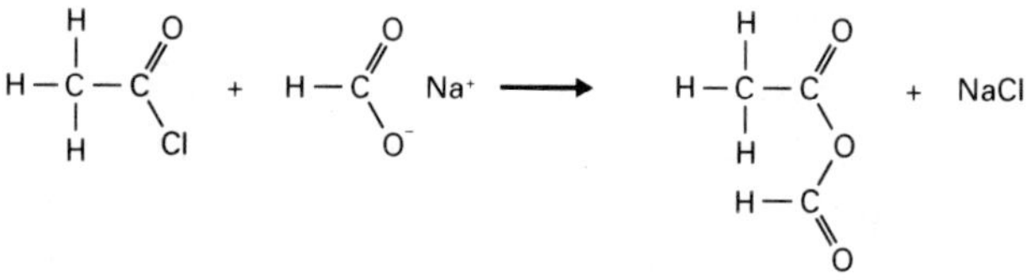

ethanoyl chloride + sodium methanoate → ethanoic methanoic anhydride + sodium chloride

Reactions of acid anhydrides

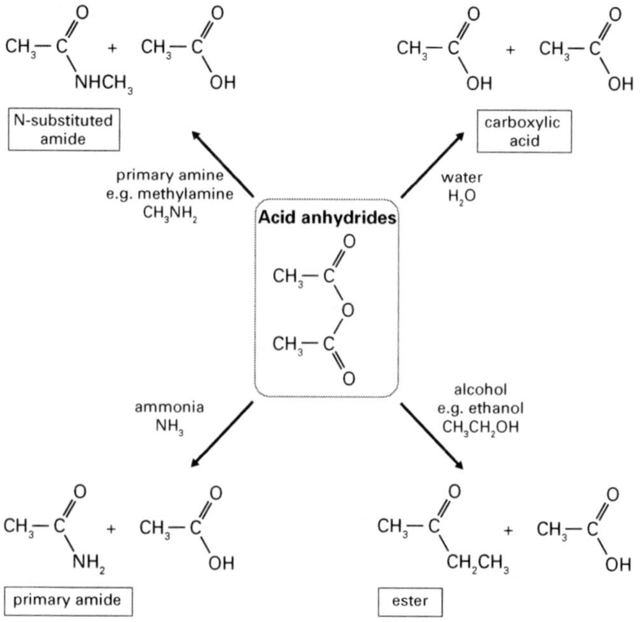

Acid anhydrides react with

- water to produce carboxylic acids
- alcohols to produce esters
- ammonia to produce primary amides
- primary amines to produce N-substituted amides (or secondary amides)

Notice that acid anhydrides produce the same primary product as acyl chlorides, but the secondary product is always a carboxylic acid instead of fumes of hydrogen chloride.

Worked example

1 Give the displayed formula and name of the products formed when propanoic anhydride is reacted with each of the following:

a ammonia

b ethanol

c ethylamine

a

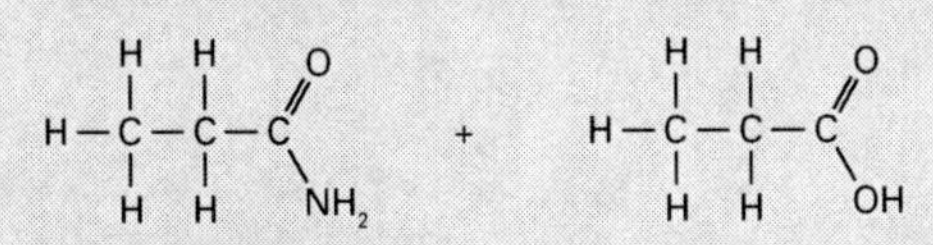

propanamide + propanoic acid

b

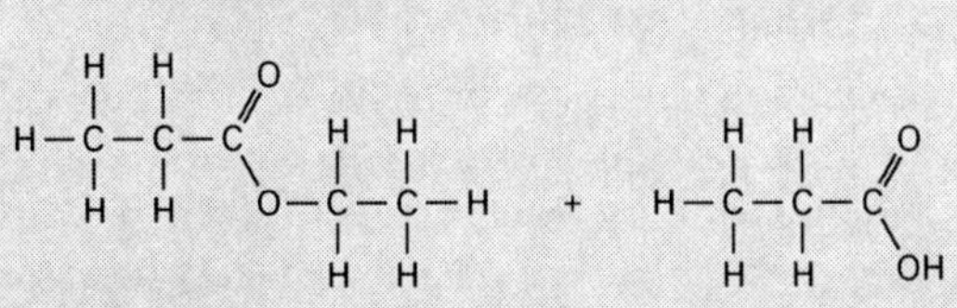

ethyl propanoate + propanoic acid

c

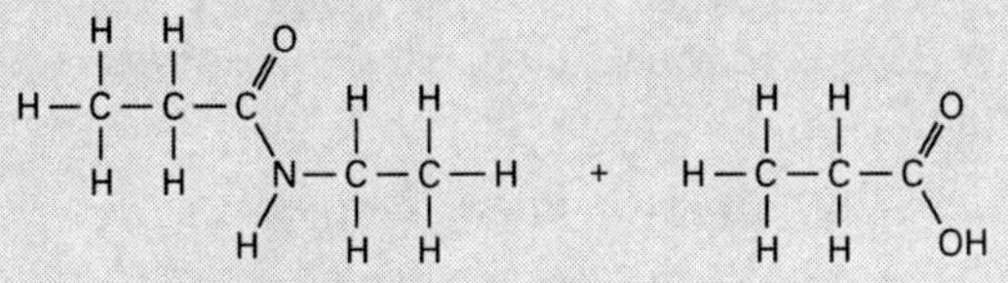

N-ethyl propanamide + propanoic acid

Aspirin

Aspirin (2-ethanoyloxybenzoic acid) is a useful medicine. It is widely used

- as a mild analgesic or painkiller
- to reduce high temperatures in fevers
- as an anti-inflammatory to reduce the swelling in conditions such as arthritis
- in low quantities as an anti-clotting agent to reduce the risk of suffering a heart attack or stroke in people with a history of cardiovascular disease

The development of aspirin

Like many medicines, aspirin is based on a naturally occurring substance, 2-hydroxybenzoic acid (salicylic acid). This used to be extracted from the bark of willow trees. People had known for thousands of years that the bark was an effective painkiller. Chemists were able to extract the active ingredient from the bark. Eventually it was possible to produce 2-hydroxybenzoic acid synthetically.

This was useful because if we have to rely on natural sources

- They may be rare or become rare or only available at certain times.
- The active ingredient may vary considerably.
- The active ingredient may be contaminated with substances which could be harmful.

Although synthetic 2-hydroxybenzoic acid was widely used as a painkiller, it produced unpleasant side effects. Chemists searched for a chemical which would have the same analgesic effects but without so many side effects. The answer lay in reacting 2-hydroxybenzoic acid with ethanoyl chloride to produce the chemical we call aspirin.

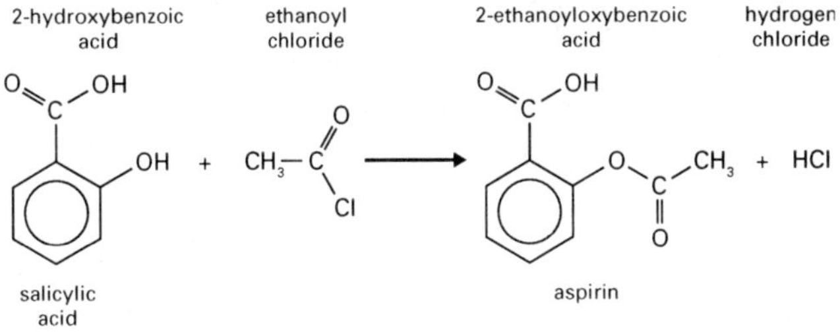

The manufacture of aspirin

Aspirin can be made by reacting 2-hydroxybenzoic acid with either the acid anhydride ethanoic anhydride or with the acyl chloride ethanoyl chloride.

Advantages of using ethanoic anhydride in the manufacture of aspirin

Although ethanoyl chloride is more reactive than ethanoic anhydride, it is ethanoic anhydride which is used in the reaction.

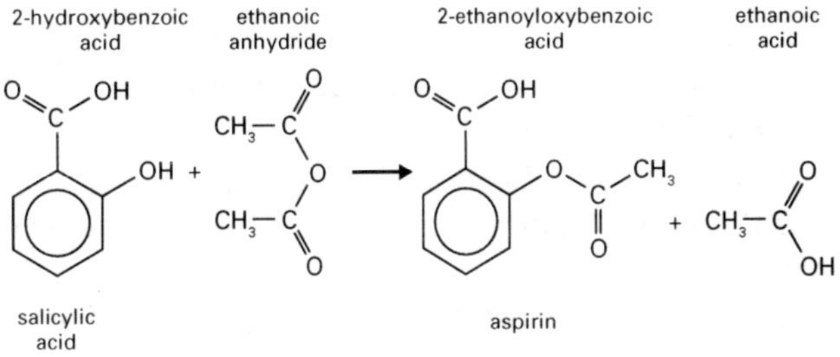

This is because ethanoic anhydride is:

- cheaper
- less reactive, so it is safer and easier to control
- less susceptible to hydrolysis
- less corrosive because it does not produce a secondary product of fumes of hydrogen chloride. The ethanoic acid which is produced is much easier to deal with.

Questions

1 Name these compounds:

a

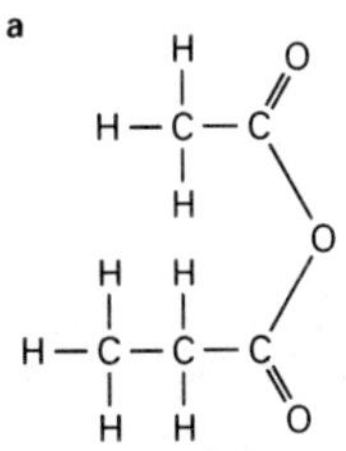

b

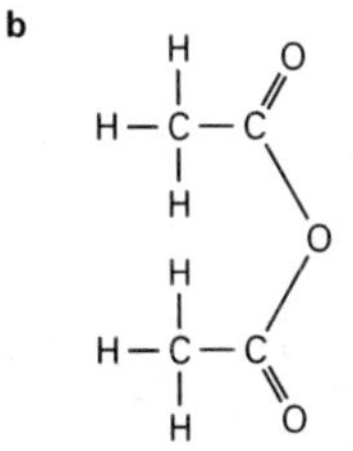

c

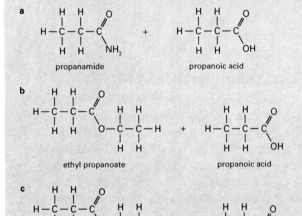

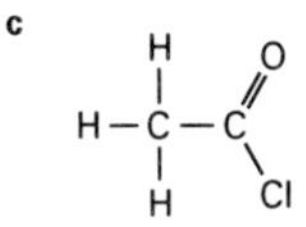

2 Suggest why acid anhydrides are more widely used than acyl chlorides during organic synthesis.

3 Name the two organic products made when ethanoic anhydride is reacted with ethanol.

4.13 Aromatic compounds 1

OBJECTIVES

By the end of this section you should know:

- *about the nature and bonding in a benzene ring*
- *that delocalization confers stability to benzene*
- *how to use evidence from enthalpies of hydrogenation to prove that delocalization confers stability*
- *that electrophilic attack of benzene results in substitution*
- *the mechanism for the nitration of benzene*

Arenes

Arene molecules contain at least one **benzene** ring. In the past arenes were called **aromatic compounds** because of their distinctive smells.

Benzene

Reminder

Benzene must not be used in schools because it is highly toxic. Repeated contact with benzene can cause leukaemia.

- Benzene has the molecular formula C_6H_6.
- It is a planar molecule with six carbon atoms arranged in a hexagonal ring.
- The bond angle between the atoms is 120°.

Ideas about the structure of benzene have developed over time.

Kékule suggested that benzene molecules contained a ring of six carbon atoms with alternating single and double bonds.

The displayed formula of the Kékule structure of benzene.

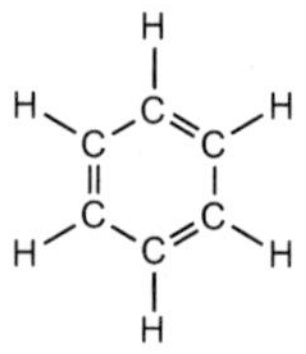

The skeletal formula of the Kékule structure of benzene.

The modern structure of benzene

All the carbon bonds are the same length.

bond	bond length (nm)
mean C–C	0.154
mean C=C	0.134
C–C bonds in benzene	0.140

- The carbon bonds in benzene are **intermediate** between single and double bonds.
- Benzene has six carbon atoms joined by single bonds with **delocalized** electrons above and below the plane.
- Each carbon atom contributes one electron to form a bond that is spread over all six carbon atoms.
- These electrons are delocalized.
- Benzene is a **planar** (flat) molecule, which allows the p orbitals to overlap.

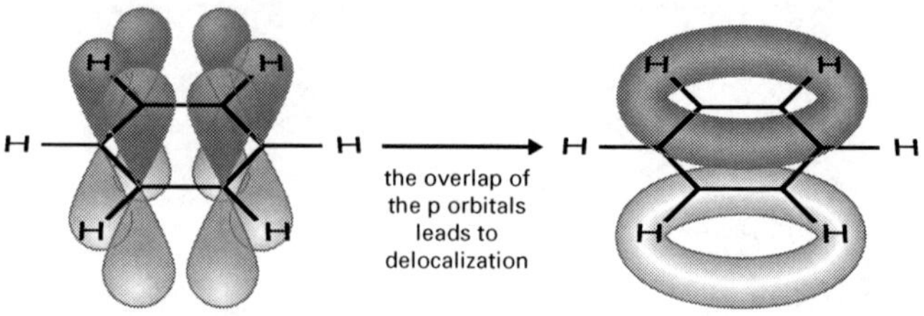

Benzene is often represented by its skeletal formula.

Delocalization and stability

The delocalization of the electrons means that benzene is more stable than might be expected. Benzene does not readily undergo addition reactions, as this would require a lot of energy to break up the delocalized electron system.

Instead most of the reactions of benzene are substitution reactions. In these reactions a new group replaces a hydrogen atom and the delocalized electron system is maintained.

The delocalization enthalpy of benzene

A cyclohexene molecule has one double bond. It reacts with hydrogen in a **hydrogenation** reaction to form cyclohexane.

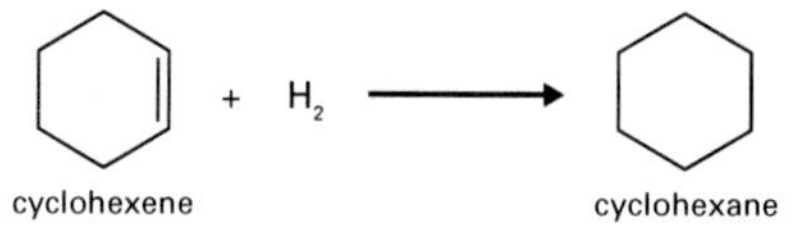

The enthalpy change for the hydrogenation of cyclohexene is –120 kJ mol^{-1}

If a benzene molecule contained alternating single and double bonds as shown by the Kékule structure it could be called cyclohexatriene.

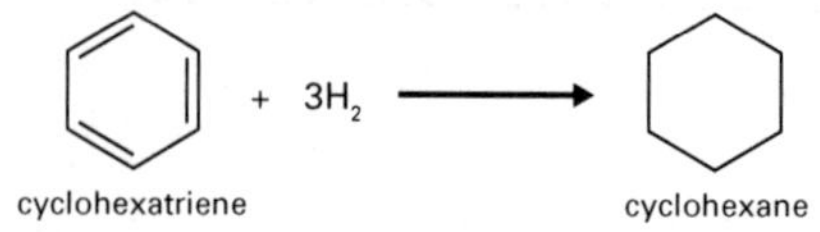

We could predict that the enthalpy change for the hydrogenation of cyclohexatriene would be equal to three times the enthalpy change for the hydrogenation of cyclohexene, which would be –360 kJ mol^{-1}

The experimental value for the enthalpy change for the hydrogenation of benzene is –208 kJ mol^{-1}

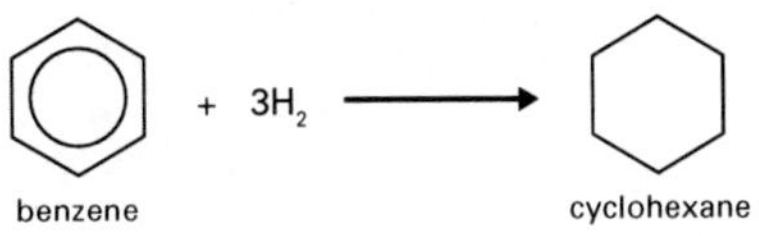

- Benzene does not contain three double bonds.
- Instead it has a delocalized electron system.
  As a result benzene is more stable than might be expected.
- More energy is required to break the delocalized electron system than is required to break three double bonds
  As a result the overall enthalpy change for the reaction is much lower than might be expected.

The difference between the predicted enthalpy of hydrogenation of cyclohexatriene and the enthalpy of hydrogenation of benzene is called the delocalization enthalpy and is equal to -152 kJ mol^{-1}

It is the amount of energy that must be supplied to break the delocalized electron system.

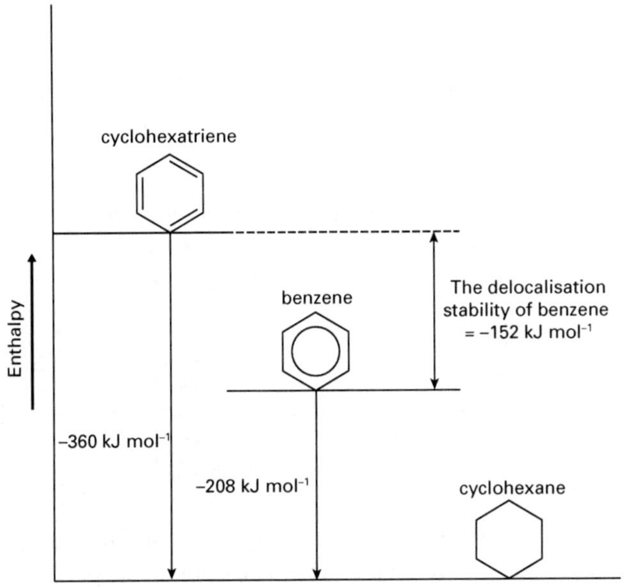

This diagram shows the delocalization enthalpy of benzene.

Electrophilic substitution

Benzene mainly undergoes **electrophilic substitution** reactions with **electrophiles** such as NO_2^+, Cl^+, and CH_3^+.

- An electrophile is a species that can accept a lone pair of electrons.
- Benzene's delocalized electron system makes it very stable.
- By undergoing substitution reactions rather than addition reactions benzene retains the stability associated with a delocalized system.

Electrophilic substitution reactions take place in two steps.

Step 1 An electrophile, E^+, is added to the benzene ring.

Step 2 A hydrogen ion is eliminated.

- In step 1 a co-ordinate bond is formed between a carbon atom in the benzene molecule and the electrophile.
- Note that the curly arrow represents the movement of a pair of electrons.
- The hexagon diagram represents a benzene molecule, C_6H_6.
- This produces an unstable intermediate which has a positive charge.
- In step 2 the bond between the carbon atom in the benzene ring and the hydrogen atom is broken. A hydrogen ion is released and the pair of electrons is used to restore the delocalized electron system.

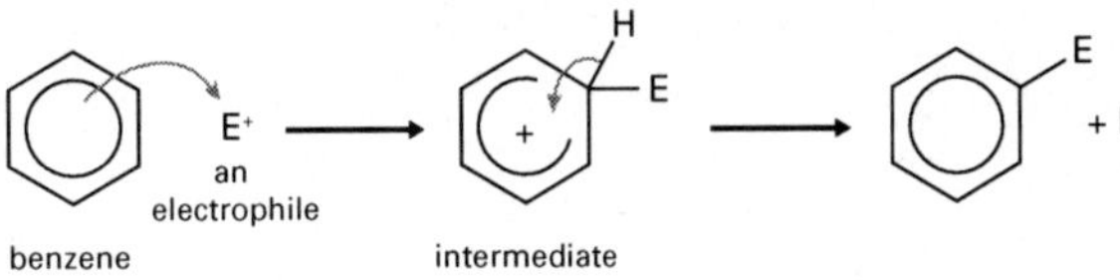

The nitration of benzene

Nitrobenzene is made by the nitration of benzene. During nitration a nitro group, NO_2, is substituted onto the benzene molecule. The overall equation for the reaction is:

$$C_6H_6 + NO_2^+ \rightarrow C_6H_5NO_2 + H^+$$

- NO_2^+ is called a **nitronium** ion. It is an electrophile, which is generated by using a nitrating mixture.
- A nitrating mixture consists of a mixture of concentrated sulfuric acid and concentrated nitric acid.
- Sulfuric acid is a stronger acid than nitric acid, so sulfuric acid will protonate the nitric acid:
  $$H_2SO_4 + HNO_3 \rightarrow HSO_4^- + H_2NO_3^+$$
- The protonated nitric acid, $H_2NO_3^+$ then decomposes to form a nitronium ion, NO_2^+:
  $$H_2NO_3^+ + H_2SO_4 \rightarrow NO_2^+ + H_3O^+ + HSO_4^-$$
- The nitronium ion electrophile then reacts with benzene.
- The reaction is carried out at about 50 °C. If higher temperatures are used further nitration may occur.

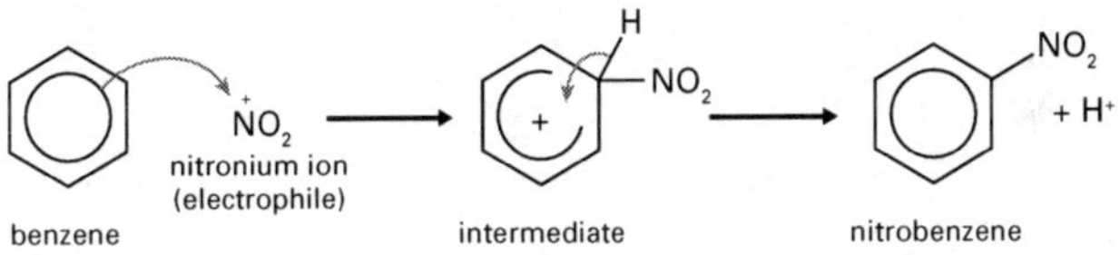

- Finally, the catalyst, sulfuric acid, is regenerated:
  $$H^+ + HSO_4^- \rightarrow H_2SO_4$$
- Overall a nitro group is added to the benzene molecule and hydrogen is lost.

Nitration is an important step in the synthesis of many useful new substances.

Questions

1 Name these compounds

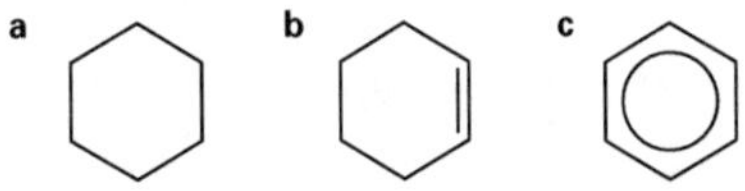

2 Why haven't you done any practical work using benzene?

3 Why does benzene typically undergo substitution reactions rather than addition reactions?

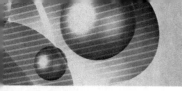

OBJECTIVES

By the end of this section you should know:

- *that nitration is an important step in synthesis, for example in the production of explosives and in the formation of amines from which dyestuffs are made*
- *that Friedel–Crafts acylation reactions are important steps in synthesis*
- *the mechanisms for the acylation of benzene and about the role of the $AlCl_3$ catalyst*

Naming arene compounds

Arenes contain at least one benzene ring. There is no easy pattern to naming arenes so you need to learn these examples.

The names of some arenes are based on a **benzene** molecule which has groups substituted onto it.

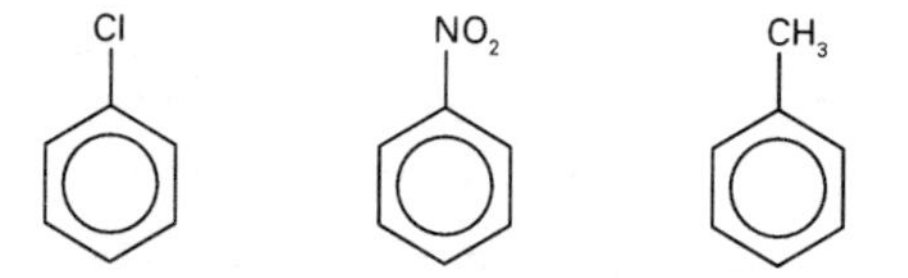

chlorobenzene   nitrobenzene   methylbenzene

The names of other arenes are based on compounds which have a **phenyl**, C_6H_5, group.

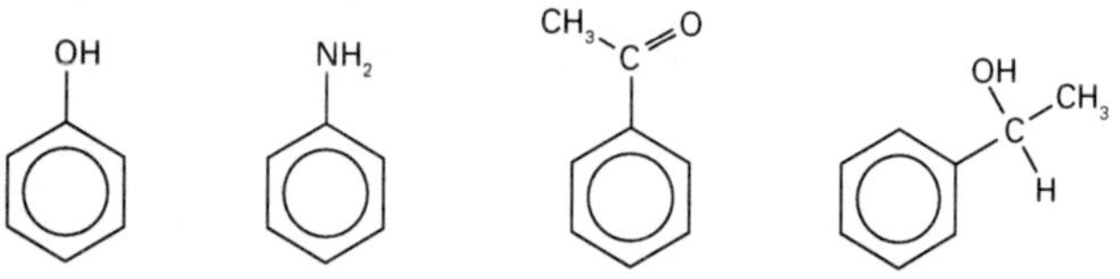

phenol   phenylamine   phenylethanone   phenylethanol

Trinitrotoluene, TNT

- 2-methyl-1,3,5-trinitrobenzene is better known as trinitrotoluene or simply **TNT**.
- TNT is a useful explosive made by the nitration of methyl benzene (toluene).

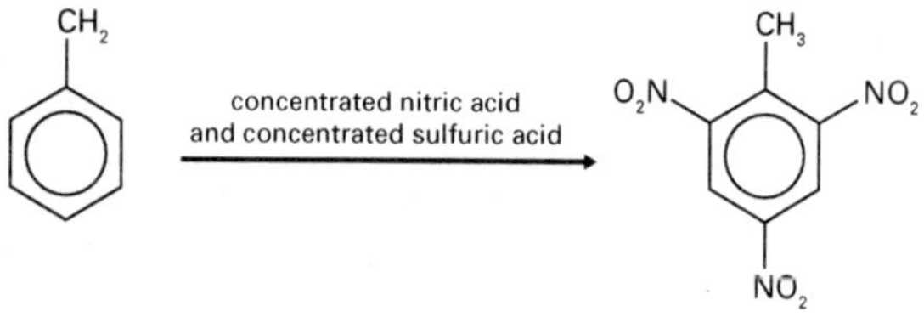

- TNT is very stable and a detonator must be used to set it off.
- It is not set off by friction or by being accidentally dropped.

  As a result TNT is relatively easy to handle compared to many other explosives. It can also be safely stored for long periods of time.

Dye stuffs

So-called azo dyes can be made by nitrating benzene to produce nitrobenzene.

- First the nitrobenzene is reduced to **phenylamine**, $C_6H_5NH_2$, an aromatic amine.

$$\underset{\text{nitrobenzene}}{C_6H_5NO_2} + 6[H] \longrightarrow \underset{\text{phenylamine}}{C_6H_5NH_2} + 2H_2O$$

- Then the aromatic amine is converted to a **diazonium salt**. Sodium nitrate(III) and hydrochloric acid are used to produce nitrous acid, HNO_2:

$$NaNO_2 + HCl \rightarrow HNO_2 + NaCl$$

- The reaction must be cooled in an ice bath so that it is kept below 10 °C. This stops the diazonium salt decomposing.

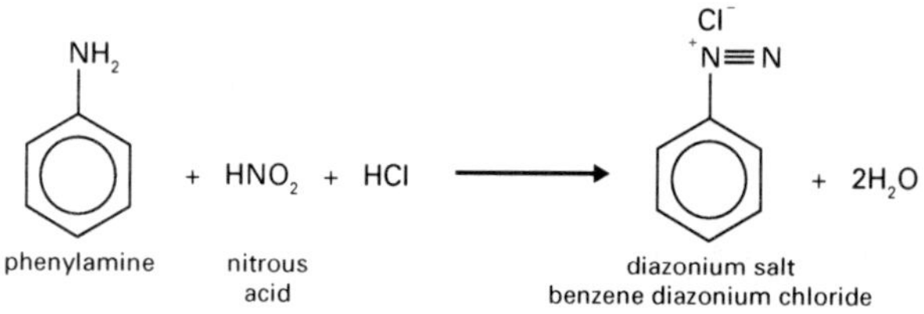

- Finally the diazonium salt is reacted with another suitable aromatic compound to form an **azo dye**.

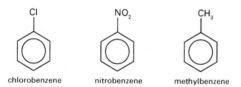

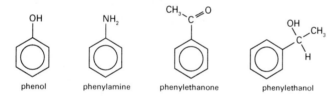

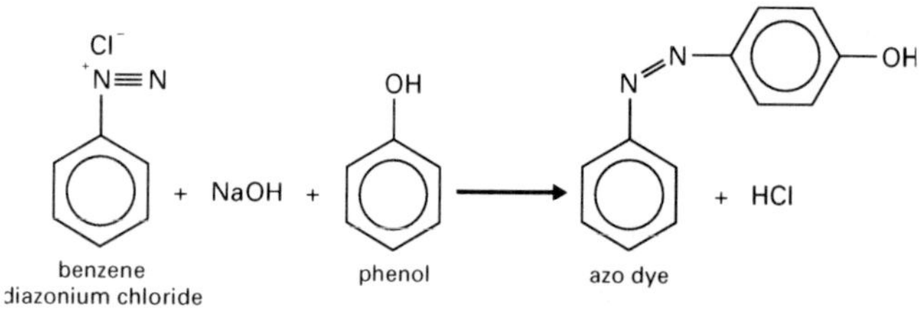

- This is a coupling reaction. Again the reaction must be cooled to stop the diazonium salt from decomposing.

Azo dyes are useful dyes that can be used to colour synthetic or natural fibres. Different coloured dyes are formed by coupling the diazonium salt with different aromatic compounds.

Worked example

Nitrobenzene, $C_6H_5NO_2$, can be reduced to form phenylamine.

1 *Give the molecular formula of phenylamine.*

Phenylamine can be used to manufacture azo dyes. Phenlyamine must first be converted to benzenediazonium chloride. The reaction mixture must be cooled in an ice bath.

2 a *What are the reagents needed for this conversion.*

b *Why must the reaction mixture be cooled?*

3 *Benzenediazonium chloride can be reacted with phenylamine to produce a yellow azo dye. Name the type of reaction which takes place between benzenediazonium chloride and phenylamine.*

(continued next column)

4 Give the displayed formula of the yellow azo dye formed in this reaction.

1 $C_6H_5NH_2$

2 a Sodium nitrate(III) and hydrochloric acid which react together to form nitrous acid

b to stop the diazonium salt from decomposing

3 It is a coupling reaction.

4

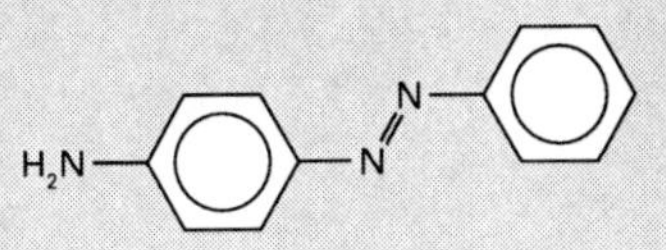

Friedel–Crafts acylation

Benzene typically undergoes **electrophilic substitution** reactions. In these reactions a new group is substituted for a hydrogen atom but the delocalized electron system is maintained.

In **Friedel–Crafts acylation** reactions benzene reacts with acyl chlorides, for example ethanoyl chloride, CH_3COCl, to form **aromatic ketones** (these are compounds that contain both a ketone group and a phenyl group).

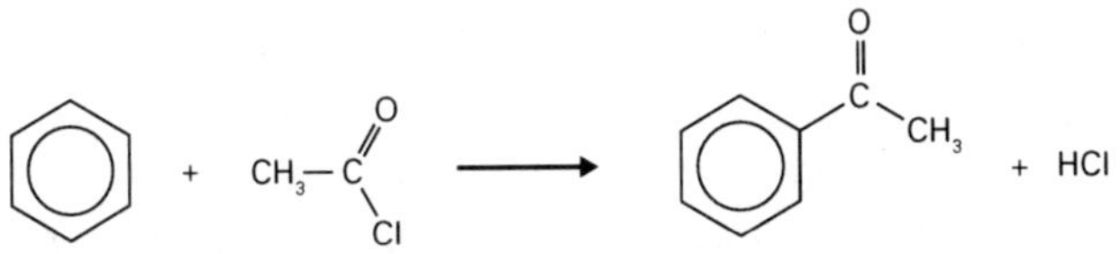

benzene ethanoyl chloride phenylethanone

These reactions must be carried out in the presence of a catalyst such as **aluminium chloride**, $AlCl_3$.

Aluminium chloride

- Aluminium chloride, $AlCl_3$, can be used as a catalyst in the acylation of benzene.
- Aluminium chloride is electron deficient. The aluminium atom has a vacant orbital, which can accept a lone pair of electrons.

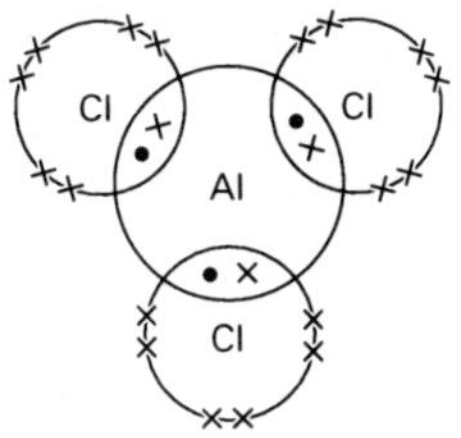

- Acyl chlorides react with aluminium chloride to produce an electrophile.

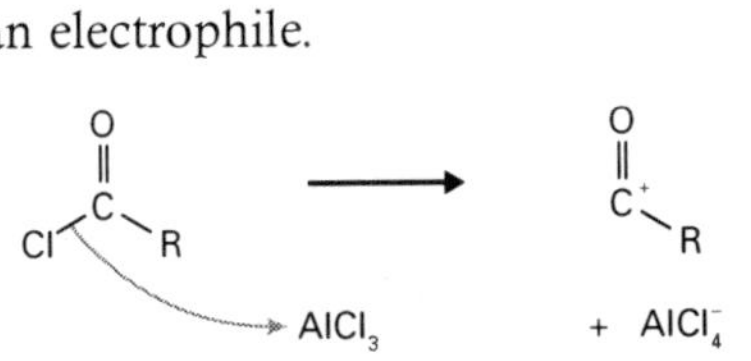

- Notice that the positive charge is written by the carbon atom in the carbonyl bond.
- The electrophile then reacts with benzene in an electrophilic substitution reaction known as a Friedel–Crafts reaction.

The mechanism of Friedel–Crafts acylation

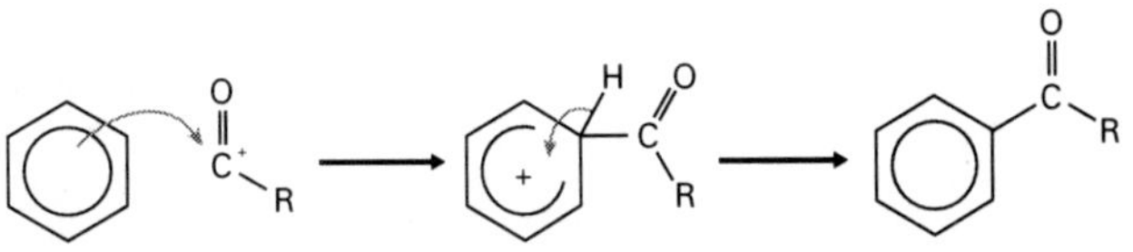

benzene intermediate aromatic ketone

- Aluminium chloride reacts with acyl chlorides to produce a tetrachloroaluminium ion $[AlCl_4]^-$ and an electrophile which goes on to react with the benzene molecule.
- The hydrogen ion lost by the benzene molecule reacts with the tetrachloroaluminium ion to regenerate the aluminium chloride, $AlCl_3$ and hydrogen chloride, HCl.

$$[AlCl_4]^- + H^+ \rightarrow AlCl_3 + HCl$$

- As phenyl ketones are less reactive than benzene, only one substitution occurs.

Worked example

Benzene reacts with ethanoyl chloride to produce an organic compound with the molecular formula C_8H_8O.

1 Name the catalyst used in this reaction and explain why it is needed.

2 Give the mechanism of the reaction.

1 The catalyst is aluminium chloride. This reacts with the ethanoyl chloride to produce the electrophile CH_3C^+O which goes on to react with the benzene molecule.

2

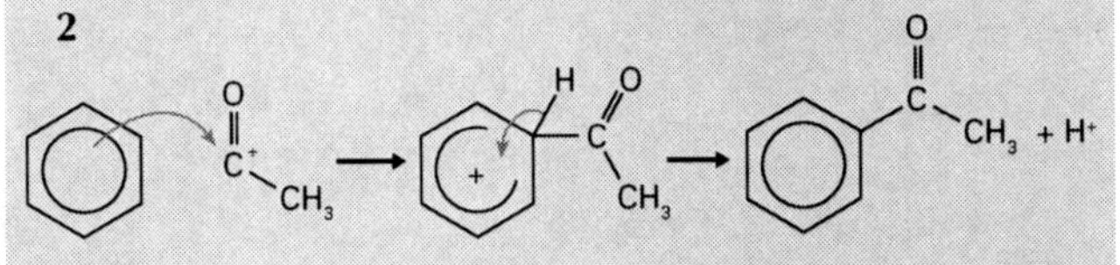

intermediate phenylethanone

The importance of Friedel–Crafts acylation reactions in synthesis

The Friedel–Crafts acylation of benzene substitutes a ketone group onto the benzene ring to produce an aromatic ketone. Changes can then be made to the group which is now attached to the benzene ring – for example the ketone group can be reduced using reducing agents such as dry $LiAlH_4$ to form an aromatic secondary alcohol.

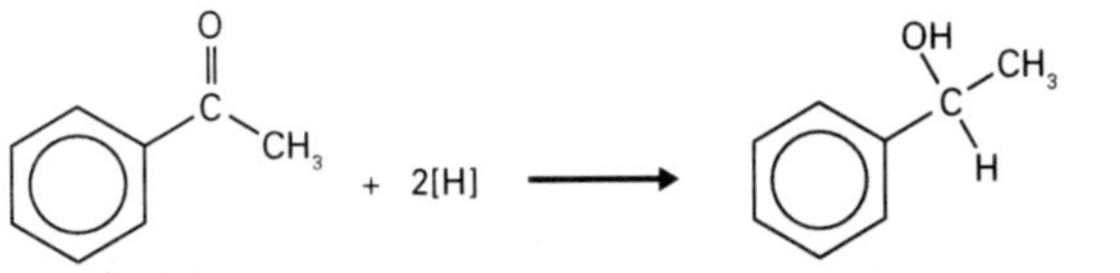

phenylethanone (aromatic ketone) phenylethanol (aromatic secondary alcohol)

Questions

1. Why is TNT a comparatively safe explosive to handle?
2. Why must diazonium salts be cooled?
3. Name the catalyst used in the Friedel–Crafts acylation of benzene.

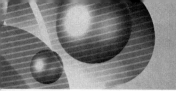

4.15 Amines and amino acids

OBJECTIVES

By the end of this section you should know:

- *the displayed formulae for primary, secondary, and tertiary amines*
- *how to name amines*
- *how to recognize the displayed formula of quaternary ammonium ions*
- *how amines can be prepared*
- *about the difference in base strength between ammonia, primary aliphatic amines, and primary aromatic amines*
- *that quaternary ammonium salts can be used as cationic surfactants*
- *that amino acids have acidic and basic properties and how they can form zwitterions*
- *how mixtures of amino acids can be separated*

Naming amines

Amines are fishy-smelling compounds which contain nitrogen atoms and are derived from ammonia, NH_3.

Amines are classified as primary, secondary, or tertiary depending on the number of hydrogen atoms in the ammonia that have been replaced by organic groups.

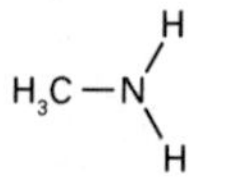
methylamine

1 hydrogen is replaced so this is a primary amine

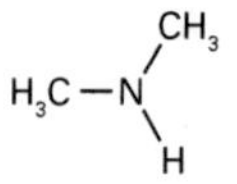
N-methylmethanamine

2 hydrogens are replaced so this is a secondary amine

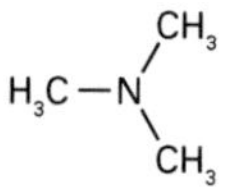
N,N-dimethylmethanamine

3 hydrogens are replaced so this is a tertiary amine

Aromatic amines contain a benzene ring.

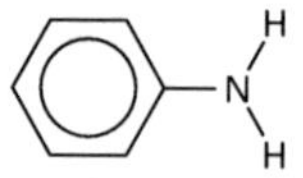
phenylamine

The primary amine propanamine exists as two positional isomers.

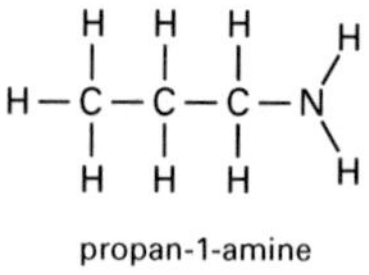
propan-1-amine

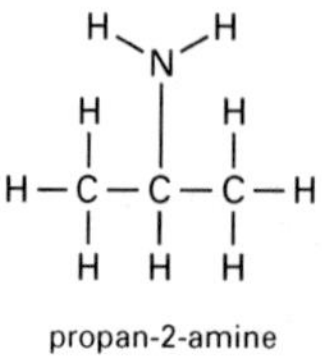
propan-2-amine

The basic properties of amines

- Ammonia, primary amines, and primary aromatic amines are **Brønsted–Lowry bases** because they can accept protons.
- The nitrogen atom in an ammonia molecule has a lone pair of electrons that can accept a hydrogen ion (or proton) to form an ammonium ion.

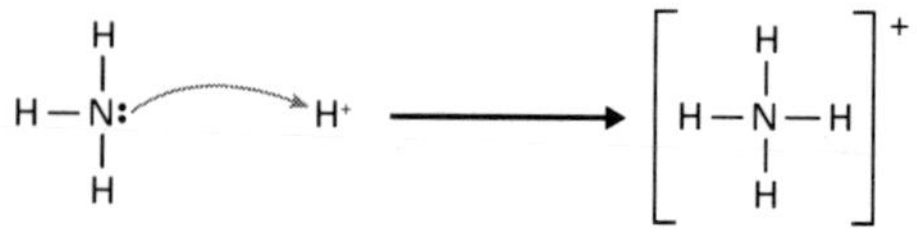

- When ammonia dissolves in water it reacts with the water molecules:
  $NH_3(aq) + H_2O(l) \rightleftharpoons NH_4^+(aq) + OH^-(aq)$
- The solution formed is alkaline because hydroxide, OH^-, ions are formed. However, the reaction does not go to completion, so ammonia is only a weak base.
- Primary amines also dissolve in water and form alkaline solutions.
- Primary amines are stronger bases than ammonia.
  $CH_3NH_2(aq) + H_2O(l) \rightleftharpoons CH_3\overset{+}{N}H_3(aq) + OH^-(aq)$
- The **alkyl** or R group increases the electron density on the nitrogen atom in the amine molecule.
- This makes the lone pair of electrons on the nitrogen more available to form dative covalent bonds. As a result primary amines are stronger bases than ammonia.

Primary aromatic amines such as phenylamine are weaker bases than ammonia.

- The lone pair of electrons on the nitrogen atom become partially delocalized onto the benzene ring.
- This reduces the electron density on the nitrogen and makes its lone pair less available to form dative covalent bonds.

Nucleophilic properties of amines

Ammonia and amines also act as **nucleophiles**. They have a lone pairs of electrons which can be donated to form a dative covalent bond.

Haloalkanes react with an excess of ammonia to form primary amines.

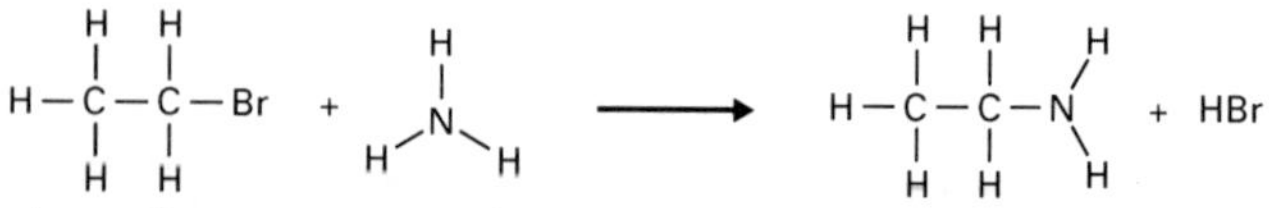
bromoethane ammonia ethanamine

- The reactivity of the haloalkane depends on the strength of the carbon–halogen bond.
- This means that fluoroalkanes are least reactive and iodoalkanes are most reactive.
- The reaction has a nucleophilic substitution mechanism.

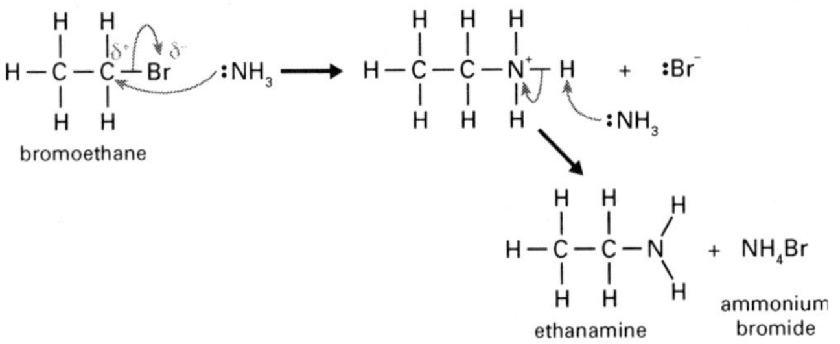

- The first ammonia molecule acts as a nucleophile, donating a lone pair of electrons to form a dative covalent bond.
- The second ammonia molecule acts as a base, accepting a proton.

However, if ammonia is not in excess the primary amine produced will react with another haloalkane molecule and further substitution can take place.

The lone pair of electrons on the nitrogen atom in the amine can acts as a nucleophile so you can end up with a mixture of a primary amine, a secondary amine, a tertiary amine, and a quaternary ammonium salt.

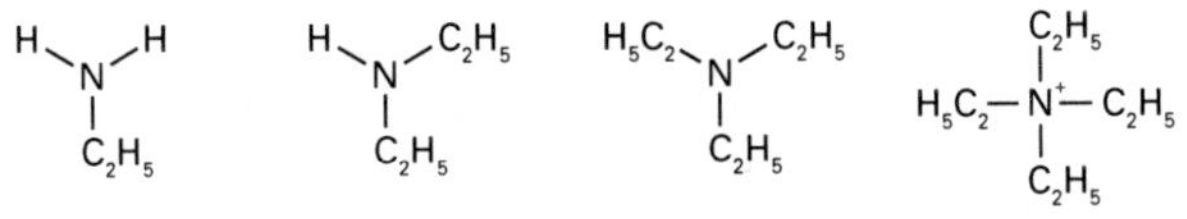

primary amine — secondary amine — tertiary amine — quaternary ammonium ion

These mixtures can be separated by fractional distillation.

Cationic surfactants

Quaternary ammonium salts are used as **cationic surfactants**. These products reduce the surface tension of liquids.

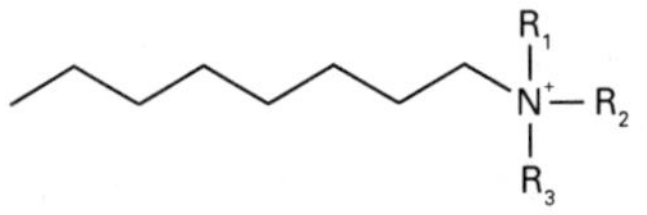

The positive charge of the ammonium ion allows them to bond with water, while the non-polar alkyl end bonds with oils and fats.

Quaternary ammonium salts are also used as ingredients in fabric conditioners and hair shampoos and conditioners. The positively charged ions bind to the fibre or hair, allowing the strands to move past each other more easily and so reducing static.

Making amines

- Primary amines can be prepared by heating haloalkanes with excess ammonia (see section on the nucleophilic properties of amines)
- They can also be prepared by the reduction of nitriles.
- In the laboratory nitriles are reacted with lithium tetrahydridoaluminate(III), $LiAlH_4$, in dry ether followed by dilute acid.
- In industry, hydrogen gas and a nickel or platinum catalyst are used.
- Aromatic amines such as phenylamine can be made by the reduction of nitro arenes such as nitrobenzene.

Amino acids

- **Amino acids** have an amino, $-NH_2$, group and a carboxyl, $-COOH$, group.
- The amino group is basic and the carboxyl group is acidic so amino acids are amphoteric – they react with both acids and bases.
- There are about twenty naturally occurring amino acids.

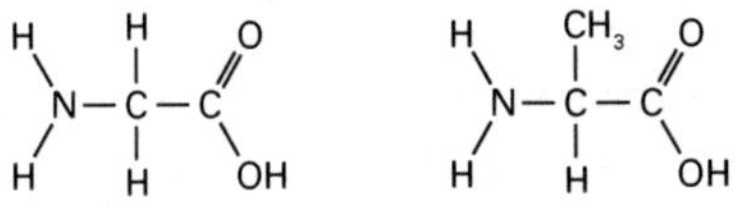

α-amino acid — alanine

- Different amino acids have different R groups.
  α-Amino acids have an –R group bonded to a carbon atom that directly carries the amino and carboxyl groups.
  α-Amino acids join together to form proteins (see also section 16 Polymers 1).

Zwitterions

- Amino acids can exist as **zwitterions** in which both the functional groups are charged.

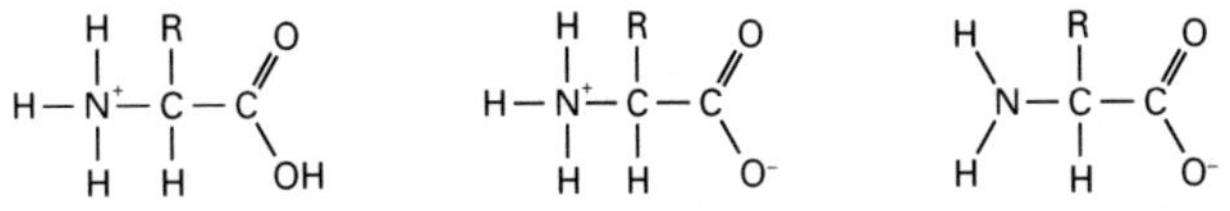

lower pH — zwitterion isoelectric point — higher pH

- The **isoelectric point** is the pH at which both the amino group and the carboxyl group are most likely to be charged.
  As a result the amino acid will be electrically neutral overall.
- The isoelectric point is different for different amino acids.
- At lower pH values the amino group is more likely to be protonated.
- At higher pH values the carboxyl group is more likely to have lost its proton.
- The existence of zwitterions means that amino acids are readily soluble in water (which is a polar solvent).
- The existence of zwitterions means that amino acids have quite high melting points and are solid at room temperature.

Separating mixtures of amino acids

- Mixtures of amino acids can be separated using paper chromatography.
- Each amino acid in the mixture moves a slightly different distance along the chromatogram paper.
- Ninhydrin solution is sprayed onto the colourless amino acids to make them visible.
- This method allows the mixture to be separated and the amino acids identified (see also section 4.19 Structural determination).

Questions

1 Why do amines act as bases?

2 Why do amines react as nucleophiles?

3 What is meant by the isoelectric point of an amino acid?

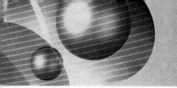

4.16 Polymers 1

OBJECTIVES

By the end of this section you should know:

- *that condensation polymers may be formed between amino acids*
- *that proteins are formed when amino acids are joined together by peptide links*
- *about the importance of hydrogen bonding in proteins*
- *that hydrolysis of the peptide links in proteins produces mixtures of amino acids*
- *that alkene molecules can be joined together to form addition polymers*
- *how to draw the repeating unit of an addition polymer from its monomer*
- *how to draw the monomer of an addition polymer from its repeating unit*

Proteins and polypeptides

Amino acids can join together to form **polypeptides** or **proteins**.

- Proteins are made up of more than about twenty different amino acids.
- Most common proteins contain more than a hundred amino acid units.
- Polypeptides are made up of less than about twenty amino acid units.

Making proteins

- Proteins are made by a **condensation polymerization** reaction.
- In these reactions, bonds form between the monomers, in this case amino acids, so that they join together to form a larger molecule.
- Another small molecule, in this case water, H_2O, is also made. As each amino acid is added one water molecule is also formed.
- Proteins are examples of natural polymers.

The peptide link

Amino acids contain the amine group, $-NH_2$, and the carboxyl group, –COOH.

The amine group of one amino acid reacts with the carboxyl group of the next amino acid to form a dipeptide. A water molecule is lost and a **peptide link**, –CONH–, is formed.

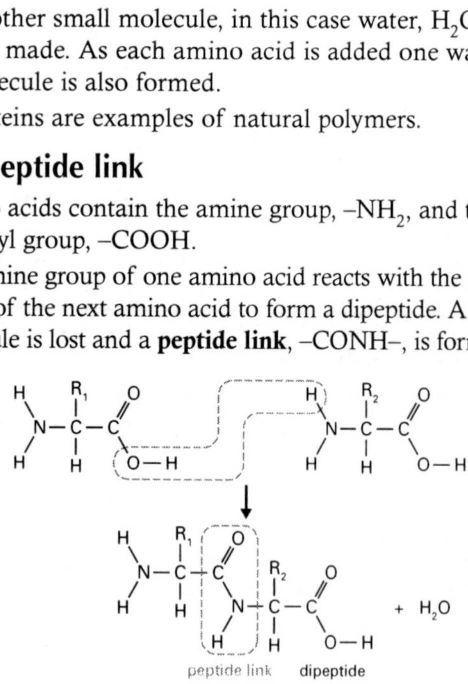

- Note the R groups may be the same or they may be different.
- In proteins the bonds named peptide bonds are really amide groups. So proteins and polypeptides are actually polyamides.
- The dipeptide produced in this reaction has an amino group at one end and a carboxyl group at the other end so the dipeptide can react with more amino acids.

Hydrogen bonding in proteins

Proteins have complicated structures. These structures are described in four layers.

- The primary structure of a protein is the order of the amino acids which make up the proteins.
- The protein chain is not a straight line. Instead it is folded into a secondary structure. The most common structure is the **α-helix**. This secondary structure is caused by hydrogen bonds between the peptide links.
- The α-helix chain is coiled or folded in a certain way to form the tertiary structure. Hydrogen bonds and other forces of attraction between different parts of the protein, such as disulfide bridges, are very important.
- Complex proteins may be formed from several separate protein chains. This is the quaternary structure of a protein.

As a result different structures have different shapes.

At high temperatures or in acidic or alkaline conditions the hydrogen bonds can be broken. This disrupts the structure of the protein and it becomes denatured.

Hydrolysis of the peptide link

- Proteins are made by the condensation polymerization of amino acids.
- Proteins can be broken down into amino acids by **hydrolysis**. Hydrolysis is the reverse of the condensation polymerization of amino acids.
- First the protein is reacted with 6 mol dm^{-3} hydrochloric acid for 24 hours.
- Then the reaction mixture is neutralized.

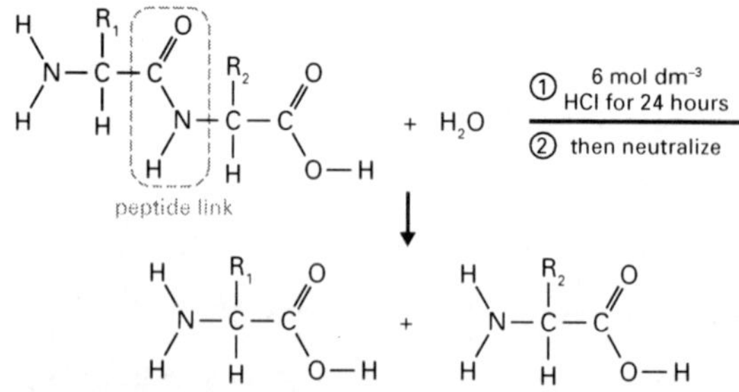

Worked example

Proteins can be split into their constituent amino acids in the laboratory by hydrolysis.

1 Give the conditions required for the hydrolysis of proteins.

2 Study the diagram of the dipeptide next column top.

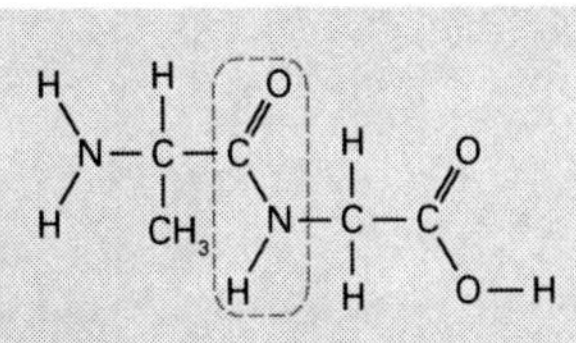

a *Name the part of the dipeptide that has been circled.*

b *Name and draw the displayed formulae of two amino acids made when this dipeptide is hydrolysed.*

1 6 mol dm^{-3} hydrochloric acid for 24 hours and then neutralize the mixture.

2 **a** peptide link

b

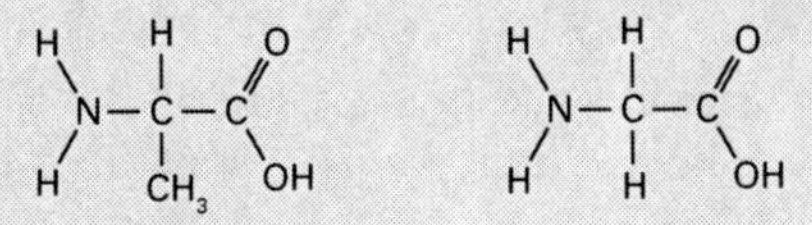

alanine glycine

Addition polymers

- Addition polymers are made when many unsaturated molecules join together.
- In **addition polymerization** the only product formed is the polymer.
- In condensation polymerization a small molecule such as water or hydrogen chloride is made at the same time as the polymer is made.

Making addition polymers

- The monomer ethene is used to produce the polymer poly(ethene) or polythene.
- The names of polymers can be written either with or without the brackets.
- Here three ethene molecules join together to form a section of poly(ethene).

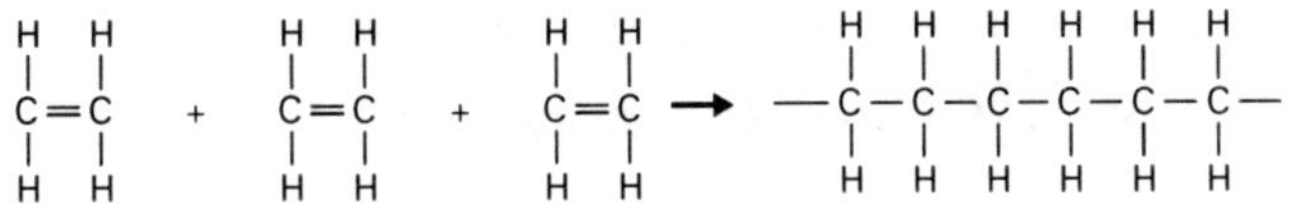

ethene molecules join together to form poly(ethene)

- Notice that the monomers are unsaturated but the polymer is saturated.
- In a similar way phenylethene molecules can join together to form poly(phenylethene).

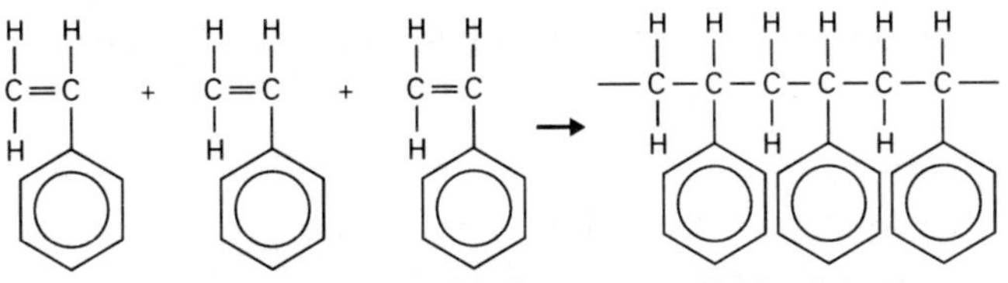

phenylethene molecules join together to form poly(phenylethene)

- Phenylethene used to be called styrene, so its polymer was called polystyrene.

The repeating unit

Because polymers can be made from enormous numbers of monomers we use the idea of a **repeating unit** to represent the structure of a polymer.

The repeating unit from the polymer

To work out the repeating unit of a polymer, for example propene

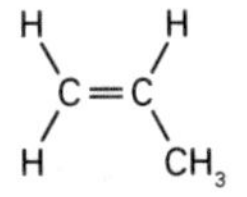

- Redraw it so that the carbon–carbon double bond is horizontal and the bond angles are changed to 90°.
- Replace the carbon–carbon double bond with a single bond and draw a line (representing a **trailing bond**) to the left and the right of the carbon atoms that had been involved in the double bond.
- Note that the brackets are not essential, but if brackets are used the trailing bonds must extend through the brackets.

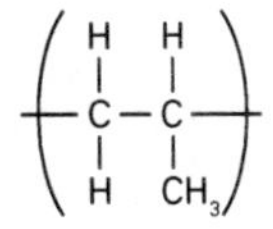

The use of the repeating unit makes it much easier to write equations for the formation of addition polymers. Here is the equation for the formation of poly(chloroethene) from its monomer chloroethene:

chloroethene poly(chloroethene)

Chloroethene used to be called vinyl chloride so its polymer was poly(vinyl chloride) or PVC.

The monomer from the repeating unit

To work out the monomer from the repeating unit of polytetrafluoroethene or PTFE

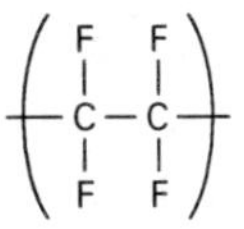

- Identify the carbon atoms which have trailing bonds.
- Remove the trailing bonds and add a double bond between these two carbon atoms.
- Finally change the bond angle from 90° to 120°.

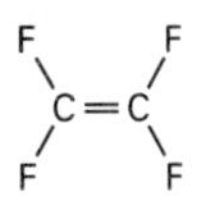

Questions

1 Identify the two functional groups found in all amino acids.

2 What are the conditions required for the hydrolysis of proteins?

3 Name the polymer formed from the addition polymerization of chloroethene.

4.17 Polymers 2

OBJECTIVES

By the end of this section you should know:

- *that dicarboxylic acids can react with diols to form condensation polymers*
- *about the linkage of the repeating units of polyesters such as Terylene*
- *that dicarboxylic acids can react with diamines to form condensation polymers*
- *about the linkage of the repeating units of polyamides such as nylon 6,6 and Kevlar*
- *that poly(alkenes) are non-biodegradable while polyesters and polyamides are broken down by hydrolysis and are therefore biodegradable*
- *about the advantages and disadvantages of different methods of disposing of polymers*

Natural and synthetic condensation polymers

In **condensation polymerization** reactions monomers join together.

- Each monomer has at least two functional groups and every time a link is made to extend the polymer a small molecule, such as water, is lost.
- Natural condensation polymers include proteins (see section 4.16 Polymers 1 and section 4.15 Amines and amino acids) and silk.
- Synthetic condensation polymers include nylon and Kevlar.

Dicarboxylic acids and diols

Dicarboxylic acids contain two carboxyl, –COOH, groups. Ethanedioic acid is an example.

- Note that even the carbon atoms involved in the carboxyl group are included in the name.
- The suffix dioic acid is used to indicate that there are two carboxyl groups.

Diols contain two hydroxyl, OH, groups.

- The carbon atoms bonded to the hydroxyl functional groups are included in the name.
- The positions of both the hydroxyl groups are also given in the name.

Propane-1,3-diol is an example.

Making polyesters

Esters can be made by reacting alcohols with carboxylic acids.

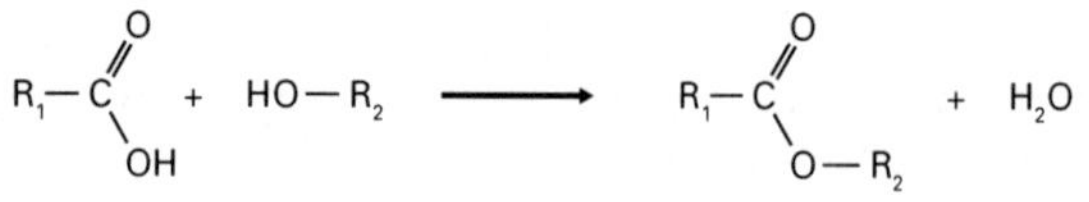

- **Polyesters** can be made by reacting dicarboxylic acids with diols.
- When a dicarboxylic acid reacts with a diol an ester and water are made.

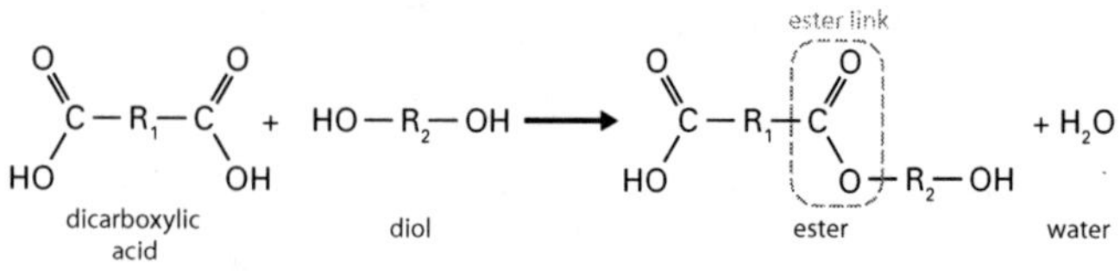

- Notice how one end of the ester still has a carboxyl group while the other end still has a hydroxyl group. As a result the ester can react with more molecules to form a long polyester molecule.

Terylene

Terylene is made from the dicarboxylic acid benzene-1, 4-dicarboxylic acid and the diol ethane-1,2-diol.

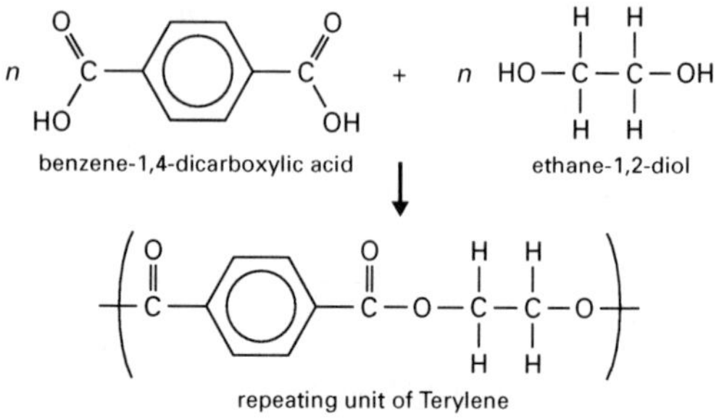

Benzene-1,4 dicarboxylic acid used to be called terephthalic acid while ethane-1,2-diol used to be called ethylene glycol.

- The polymer is called polyethylene terephthalate or PET.
- It is used to make plastic drinks bottles.
- It can also be drawn into fibres which are known as Terylene which is used in fabrics that are used to make clothes.
- The inclusion of Terylene makes the fabric less likely to crease and more likely to last longer.

The repeating unit of a polyester

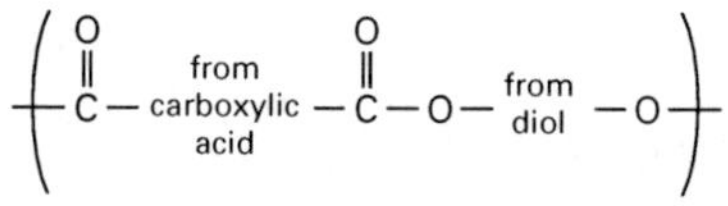

- The trailing bonds go from the carbon atom of the C=O group of the carboxylic acid on one side and from the oxygen atom of the diol group on the other side.
- If brackets are used the trailing bonds should extend through the brackets.

Diamines

Diamines have two amine, $-NH_2$, groups.

Propane-1,3 diamine is an example.

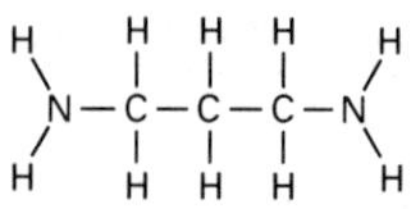

- Note that the position of both functional groups is given.
- The suffix diamine is used to indicate that there are two amine groups.

Proteins

- Amino acids can join together to form proteins (see also section 4.16 Polymers 1).
- Amino acids have at least two different functional groups. One is an amine, $-NH_2$, and the other is a carboxyl, $-COOH$, group.
- Amino acids can be described as being difunctional compounds.

Making synthetic polyamides

- An amide link is made between a molecule which has a carboxyl group and another molecule which has an amine group.

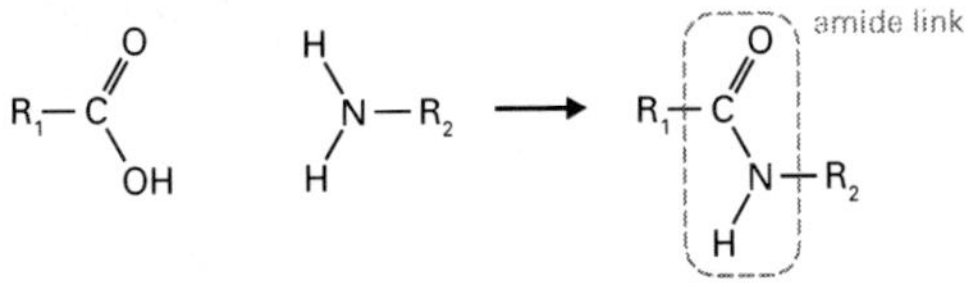

- Polyamides are made by reacting dicarboxylic acids with diamines.

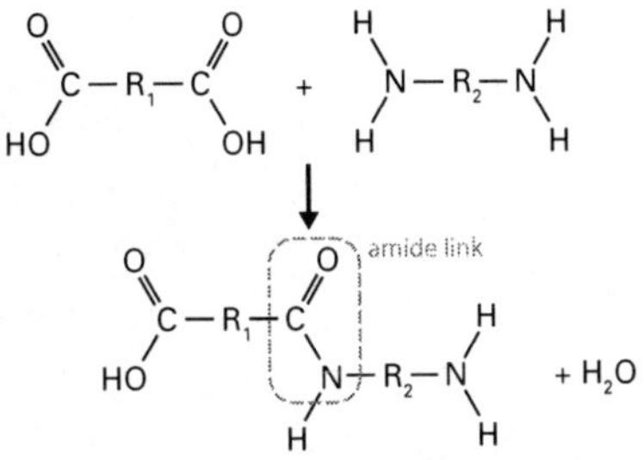

- As these molecules have a functional group at each end of the molecule they can form very long polymer chains.

The repeating unit of a polyamide

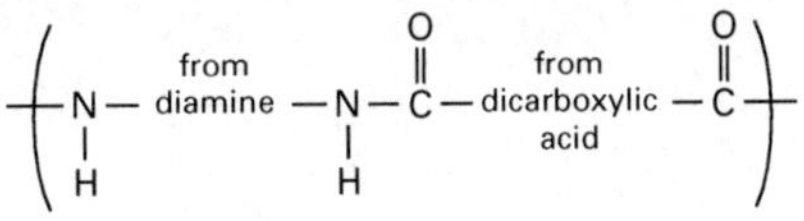

- The trailing bonds go from the nitrogen atom of the amine on one side and from the carbon atom of the C=O group of the carboxylic acid on the other side.
- If brackets are used the trailing bonds should extend through the brackets.

Nylon

- **Nylon** 6,6 is a synthetic **polyamide** used to make ropes and clothes.
- It is made from the diamine 1,6-diaminohexane and hexane-1,6-dicarboxylic acid.
- The first part of "6,6" in the name comes from the number of carbon atoms in the diamine and the second comes from the number of carbon atoms in the dicarboxylic acid.

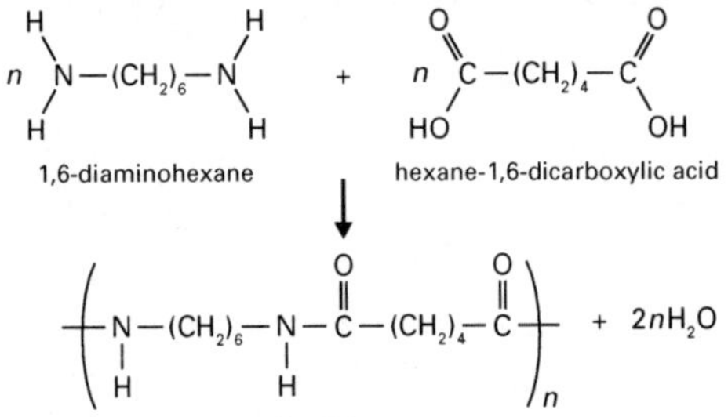

nylon 6,6

Kevlar

Kevlar is another useful synthetic polyamide.

- It is used to make bullet-proof vests.
- It is light, very strong, and has a high melting point.
- It is made by a condensation polymerization between benzene-1,4-diamine and benzene-1,4-dicarboxylic acid.

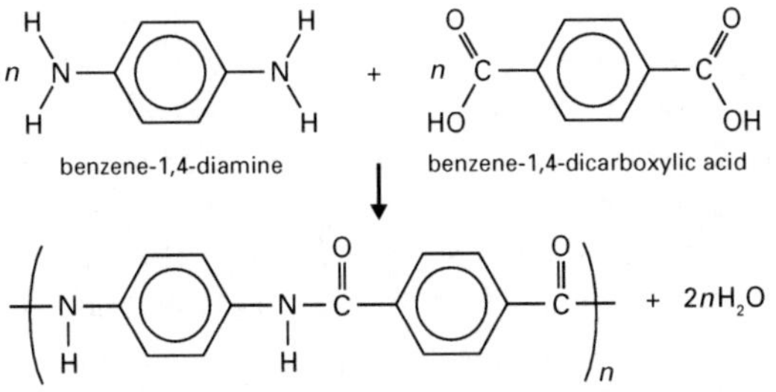

kevlar

Biodegradable and non-biodegradable plastics

- **Biodegradable** materials can be broken down by micro-organisms.
- Polyalkenes such as poly(ethene) and poly(propene) are made by addition polymerization. Addition polymers are saturated. This makes them chemically inert. As a result these plastics are **non-biodegradable**.
- Polyesters and polyamides are biodegradable and can be hydrolysed by acids or alkalis. As a result it is much easier to dispose of these plastics.

Disposing of plastics

- Plastics can be disposed of in landfill sites. Waste materials, and associated traffic can cause problems for local people.
- Plastics can be burnt. This reaction releases heat energy and the waste gas carbon dioxide; it may also produce poisonous fumes.
- Some plastic can be reused (after sorting/cleaning). This saves energy and raw materials and reduces pollution.
- Polymers can be recycled after shredding/granulating. The plastics produced may not be suitable for all uses.

Questions

1. Name the type of compound that can be reacted with a dicarboxylic acid to form a polyester.
2. Name a synthetic polyamide used to make ropes and clothes.
3. Name a natural polyamide made from amino acids.

4.18 Applications

OBJECTIVES

By the end of this section you should know:

- *that organic compounds can be converted into new organic compounds by chemical reactions*
- *how to go about synthesizing organic compounds using reactions in the specification*
- *how to identify functional groups using reactions in the specification*

Synthesis

Chemists are interested in making new chemicals from existing ones. In your examination you may be asked to suggest a way to synthesize a new compound.

This will involve you showing a good understanding of the chemistry of the range of organic compounds covered in Unit 2 of the AS course and Unit 4 of the A2 course. So it is important to revise these sections very thoroughly.

There are often several ways to produce a certain chemical so chemists choose routes which

- don't involve too many steps, as the more steps there are the lower the yield of the desired product is likely to be.
- keep costs down by not using expensive starting materials or expensive techniques
- are less likely to result in competing reactions that will decrease the yield of the desired product.

Safety also needs to be considered. Think about the specific hazards of the chemicals involved, for example they may be volatile, flammable, toxic by inhalation, or corrosive.

Then think about steps that could sensibly be used to reduce these hazards, for example

- You could heat a volatile flammable liquid under reflux using an electrical heater.
- You could wear gloves when using a corrosive chemical.
- You could use a fume cupboard when using a chemical that should not be breathed in.
- It will be assumed that a lab coat and goggles will be worn.

Worked example 1

Consider the series of reaction below.

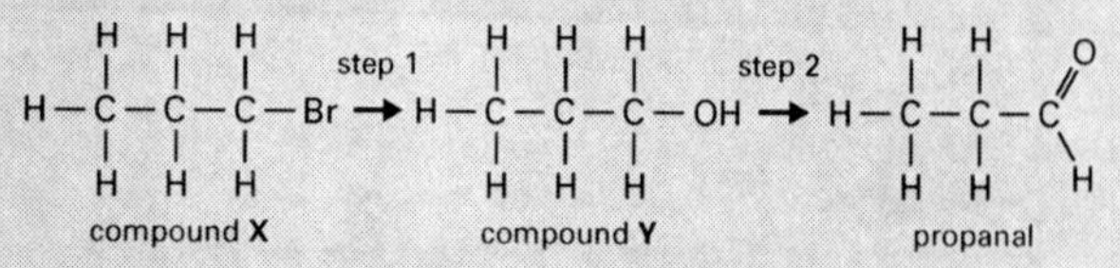

1 *Name compound **X** and compound **Y**.*

2 *Give the reagents and conditions required for*

a *Step 1*

b *Step 2*

1 **X** is 1-bromopropane. **Y** is propan-1-ol

(continues next column)

2 a Step 1 KOH(aq) and warm

b Step 2 acidified potassium dichromate(VI) and distil off the product

Worked example 2

Consider the following reaction series.

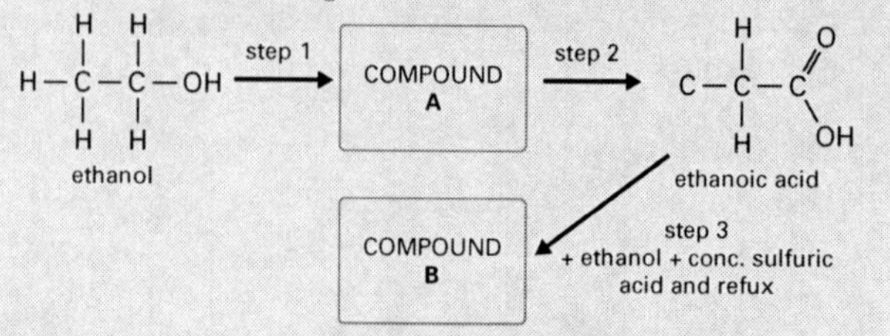

1 *Name compounds **A** and **B**.*

2 *Name the type of reaction occurring in*

a *step 1*

b *step 3*

3 *Give the reagents and conditions required for step 2 and describe what you would see.*

1 Compound **A** is ethanal and compound **B** is ethyl ethanoate.

2 a Step 1 is oxidation.

b Step 3 is esterification.

3 Warm with acidified potassium dichromate(VI) solution. There would be a colour change of orange to green.

Worked example 3

Suggest how propanoyl chloride can be made from propanal. Include the reagents and conditions that should be used.

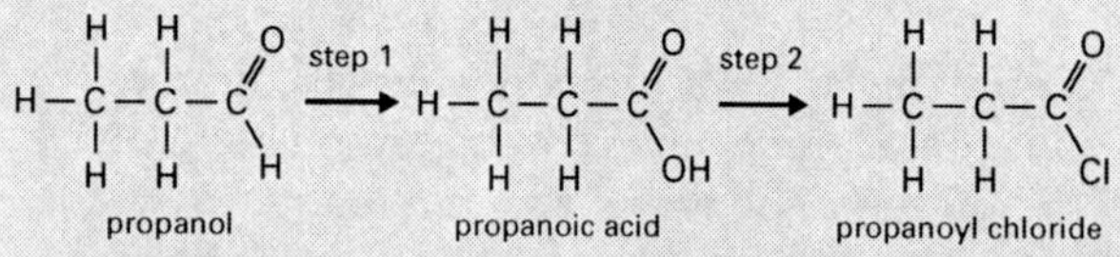

In step 1 propanal is oxidized to propanoic acid by heating with the oxidizing agent acidified potassium dichromate(VI). In step 2 the propanoic acid is reacted with phosphorus(V) chloride, PCl_5, at room temperature.

Identifying functional groups

Chemical tests can be used to identify a range of **functional groups**.

Identifying alkenes

- Aqueous bromine can be used to test for the presence of **alkenes**.
- If an alkene is present the aqueous bromine is decolorized (brown to colourless) because of an addition reaction between the alkene and the bromine.

- **Alkanes** are saturated so there is no change when alkanes are mixed with aqueous bromine.
- Benzene also does not react because the high stability of the delocalized electron system means it is resistant to addition reactions.

Identifying aldehydes and ketones

Aldehydes and **ketones** both contain a carbonyl, >CO, group. Aldehydes can be oxidized to carboxylic acids while ketones are not easily oxidized. This can be used to distinguish between these two groups.

Tollens' reagent

- Tollens' reagent is a solution of silver nitrate in aqueous ammonia. Tollens' reagent contains silver, Ag^+, ions.
- Aldehydes reduce these silver ions to silver atoms:

$$Ag^+(aq) + e^- \rightarrow Ag(s)$$

- A silver mirror is produced inside the test tube.
- There is no change for ketones.

Fehling's solution

- Fehling's solution is a blue solution which contains Cu^{2+} ions in aqueous alkali.
- Aldehydes reduce the Cu^{2+} ions to a brick-red precipitate of Cu_2O:

$$2Cu^{2+}(aq) + 2e^- + 2OH^-(aq) \rightarrow Cu_2O(s) + H_2O(l)$$

- There is no change for ketones.

Acidified potassium dichromate(VI)

- Acidified potassium dichromate(VI) can also be used to distinguish between aldehydes and ketones. Aldehydes reduce the orange $Cr_2O_7^{2-}$ ions to green Cr^{3+} ions.
- There is no change for ketones.

Identifying carboxylic acids

Carboxylic acids are weak organic acids that contain the carboxyl group, –COOH.

- Carboxylic acids have a pH of around 3 which means that they turn universal indicator orange/red.
- Carboxylic acids react with carbonates to form a salt, water, and carbon dioxide. For example ethanoic acid reacts with calcium carbonate:

$$2CH_3COOH(aq) + CaCO_3(s) \rightarrow (CH_3COO)_2Ca(aq) + CO_2(g) + H_2O(l)$$

- To confirm that the gas produced is carbon dioxide it can be bubbled through limewater. If the gas is carbon dioxide it will turn the limewater cloudy.

Identifying alcohols

Alcohols contain the hydroxyl, –OH, group.

These compounds can be classified as being primary, secondary, or tertiary alcohols.

- First the alcohol is heated with acidified potassium dichromate(VI).
- Tertiary alcohols are not easily oxidized so when they are heated with acidified potassium dichromate(VI) there is no change.
- Primary and secondary alcohols are both oxidized so when they are heated with acidified potassium dichromate(VI) the orange $Cr_2O_7^{2-}$ ions are reduced to green Cr^{3+} ions.
- Primary alcohols are oxidized to aldehydes, while secondary alcohols are oxidized to ketones. Tollens' reagent, Fehling's solution, or acidified potassium dichromate(VI) can be used to distinguish between these two types of compound.

Identifying haloalkanes

- First the **haloalkane** is warmed with aqueous sodium hydroxide. This cleaves the C–Hal bond and releases halide ions.
- Dilute nitric acid is then added to acidify the mixture.
- Finally an aqueous solution of silver nitrate is added.

If a fluoroalkane was used there would be no precipitate.

If a chloroalkane was used there would be a white precipitate of silver chloride.

If a bromoalkane was used there would be a cream precipitate of silver bromide.

If an iodoalkane was used there would be a yellow precipitate of silver iodide.

Solutions of dilute and concentrated ammonia can be used to confirm the identity of the haloalkane.

precipitate	dilute ammonia is added	concentrated ammonia is added
silver chloride	precipitate dissolves to form a colourless solution	precipitate dissolves to form a colourless solution
silver bromide	No change	precipitate dissolves to form a colourless solution
silver iodide	No change	No change

Questions

1 Suggest a safety precaution that could be used when using concentrated sulfuric acid.

2 How could a tertiary alcohol be distinguished from a primary or secondary alcohol?

3 A haloalkane is heated with sodium hydroxide solution and the mixture is then acidified using nitric acid. Aqueous silver nitrate is then added and a cream precipitate is formed. What is the name of this precipitate and how would you confirm that you are correct?

4.19 Structural determination

OBJECTIVES

By the end of this section you should know:

- *how the fragmentation of organic compounds during mass spectrometry can reveal information about the structure of the molecule*
- *how to interpret infrared spectra*
- *how to interpret n.m.r. spectra*
- *why TMS is used as a standard*
- *that column chromatography depends on the balance between solubility in the moving phase and retention in the stationary phase*
- *that gas–liquid chromatography can be used to separate mixtures of volatile liquids*

Mass spectrometry

Mass spectrometry can be used to identify organic compounds. When an organic compound is placed in a mass spectrometer it is bombarded by high-energy electrons and different electrons are knocked out of the organic molecule. This is called ionization.

- When a sample is bombarded with electrons it forms a cation with an unpaired electron, $M^{+\bullet}$. This species is a free radical.
- The peak on the mass spectrum with the highest mass/charge ratio is called the molecular ion peak or M.
- The highest peak is called the base peak.
- The more stable the cation (positive ion) the higher its peak will be.
- The molecular ion may also fragment, to produce ions with lower mass/charge ratios.

Reminder

The molecular ion is the one with the highest mass/charge ratio, it is not necessarily the one with the highest peak.

$$M^{+\bullet} \longrightarrow X^+ + Y^\bullet$$

X^+: this fragment is shown in the mass spectrum

$Y^\bullet$: this fragment has no charge so it is not shown in the mass spectrum

Fragmentation patterns

- The more stable the X^+ species the higher the peaks it produces in the mass spectrum.
- **Acylium** ions, RCO^+ and **carbocations** are more stable and so produce higher peaks than other, less stable species.
- Tertiary carbocations are more stable than secondary carbocations which are more stable than primary carbocations

  As a result, tertiary carbocations will give the highest peaks while primary carbocations will give smaller peaks.
- These fragmentation patterns allow chemists to work out information about the structure of an organic compound.

Infrared spectroscopy

- **Infrared** (IR) spectroscopy is usually used to identify the functional groups in organic compounds.
- Different bonds absorb infrared radiation with slightly different wavenumbers.
- By examining the wavenumbers of the radiation that have been absorbed by a compound we can work out which bonds must be present.

Reminder

Information about the wavenumber of radiation which a particular bond absorbs will be provided in an exam question.

- Transmittance–wavenumber graphs are used to show the wavenumber of the radiation that has been absorbed.
- Concentrate on the strong peaks where transmission falls below 70%.
- Note that O–H and N–H bonds produce very broad peaks due to hydrogen bonding.
- The section between 400 cm^{-1} and 1500 cm^{-1} is known as the **fingerprint region**.
- The fingerprint region is unique for every compound.

  As a result, if the fingerprint region of an unknown sample is found to be identical to the fingerprint of a known compound the unknown sample can be identified.

Nuclear magnetic resonance spectroscopy

Nuclear magnetic resonance (n.m.r.) spectroscopy is a powerful technique which gives information about the position of ^1H or ^{13}C atoms in a molecule.

The chemical shift

- If atoms have an odd mass number such as ^1H or ^{13}C then their nuclei will have spin. This spin can be detected using radio frequencies.
- Depending on the molecular environment around the ^1H or ^{13}C, their precise resonance frequency will vary slightly.
- This slight variation can be used to give information about the structure of a compound.
- To allow chemists to compare the spectra produced on different machines, a reference compound is added to the sample.
- The differences between the resonance frequencies of the ^1H or ^{13}C atoms in the sample and in the reference are measured. These are known as the shifts.

The **chemical shift**, δ, is given by

$$\delta = \frac{\text{shift (hertz)}}{\text{spectrometer frequency (MHz)}}$$

Note that the chemical shift is measured in parts per million (ppm)

The reference standard

Tetramethylsilane, TMS, is used as the reference standard for n.m.r. TMS is chosen because

$$CH_3-\underset{\underset{CH_3}{|}}{\overset{\overset{CH_3}{|}}{Si}}-CH_3$$

- It is fairly cheap.
- It does not react with the sample being analysed.
- All the carbon atoms are equivalent so it only produces one strong peak, which is usually upfield from the peaks produced by the other ^1H or ^{13}C nuclei.

Integrated spectra

^{13}C generally gives clearer spectra than ^1H n.m.r. This is because in most organic compounds the carbon atoms have fewer chemical environments than the hydrogen atoms. For example in ethanol.

$$^cH-\underset{\underset{^cH}{|}}{\overset{\overset{^cH}{|}}{C}}\overset{}{_2}-\underset{\underset{^bH}{|}}{\overset{\overset{^bH}{|}}{C}}\overset{}{_1}-OH^a$$

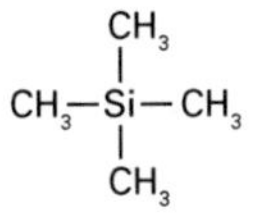

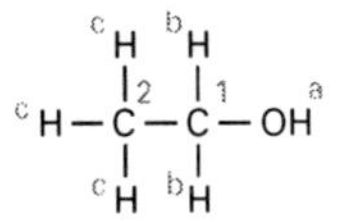

The two carbon atoms are in different chemical environments. These are labelled 1 and 2.

The ^{13}C n.m.r. **integrated spectrum** of ethanol gives two peaks of equal height.

The six hydrogen atoms in ethanol are in three different chemical environments.

- There is one hydrogen atom in the first environment. This is labelled a.
- There are two hydrogen atoms in the next environment. These are labelled b.
- There are three hydrogen atoms in the third environment. These are labelled c.

The ^1H n.m.r. spectrum of ethanol gives three peaks.

- The peaks are different heights.
- The area under the peaks in the spectra indicates the relative number of carbon atoms in the different chemical environments.

  ⊗ As a result the ratio of the areas of the peaks is 1:2:3 because this is the ratio of $H_a : H_b : H_c$.

Choosing a solvent

- Care must be taken when choosing a solvent for ^1H n.m.r. as many solvents that contain protons will produce peaks in the spectrum.
- Tetrachloromethane, CCl_4, is often used.

Spin–spin splitting patterns

The hydrogen atoms on one carbon atom affect the hydrogen atoms on the next (the adjacent) carbon atom. So in our ethanol molecule the hydrogen atoms labelled H_c influence the hydrogen atoms labelled H_b. This happens throughout the molecule and is called **spin–spin coupling**. It leads to peak splitting. This can be seen using high-resolution n.m.r.

- The $n + 1$ rule is used to deduce the spin coupling pattern, where n is the number of hydrogen atoms on an adjacent carbon atom.

- Each of the peaks in the high-resolution n.m.r. spectrum of ethanol will be affected differently.
  - H_a is not attached to a carbon atom so this peak does not undergo spin–spin coupling and remains a single peak.
  - The H_b atoms are attached to a carbon atom that is adjacent to a carbon atom which is bonded to three hydrogen atoms. This peak is split into $(3 + 1) = 4$ peaks. This is called a quartet.
  - The H_c atoms are attached to a carbon atom that is adjacent to a carbon atom which is bonded to two hydrogen atoms. This peak is split into $(2 + 1) = 3$ peaks. This is called a triplet.
- Note: a single hydrogen atom bonded to an adjacent carbon atom would split a peak into a doublet.

Gas–liquid chromatography

Gas–liquid chromatography (GLC) is used to separate mixtures of volatile liquids. In GLC, the components move through the apparatus at different speeds. Once separated the components in the mixture can be identified. Gas–liquid chromatography is so accurate that the area under the peaks indicates the relative amount of each component in a sample. It is used to measure the level of alcohol in blood or urine samples in court cases.

Column chromatography

Column chromatography can be used to separate mixtures.

- In column chromatography the stationary phase is a thin layer of a solid such as powdered silica.
- The mobile phase is the solvent.
- The mobile phase moves through the column carrying the dissolved solutes with it.
- How quickly each component in the mixture moves through the column depends on a combination of factors:
  - how soluble the component is in the solvent
  - how strongly the component is attracted to the stationary phase

Questions

1. The diagram shows a molecule of propane.

$$H-\underset{\underset{H}{|}}{\overset{\overset{H}{|}}{C}}-\underset{\underset{H}{|}}{\overset{\overset{H}{|}}{C}}-\underset{\underset{H}{|}}{\overset{\overset{H}{|}}{C}}-H$$

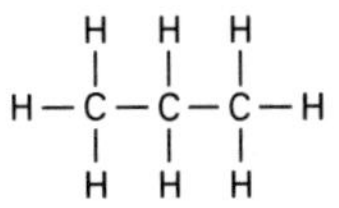

Suggest what the ^{13}C n.m.r. spectrum of propane would look like.

2. The diagram shows a molecule of propanoic acid

$$H-\underset{\underset{H}{|}}{\overset{\overset{H}{|}}{C}}-\underset{\underset{H}{|}}{\overset{\overset{H}{|}}{C}}-\overset{\overset{O}{\parallel}}{C}\diagdown_{O-H}$$

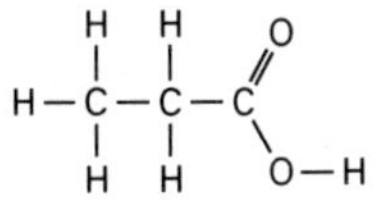

Suggest what the low-resolution ^1H n.m.r. spectrum of propanoic acid would look like.

3. Suggest what the high-resolution ^1H n.m.r. spectrum of propanoic acid would look like.

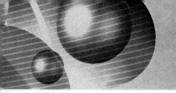

5.01 Enthalpy changes

OBJECTIVES

By the end of this section you should be able to:

- *understand that reactions can be exothermic or endothermic*
- *define standard enthalpy of formation*
- *define Hess's law*
- *define bond dissociation enthalpy*
- *calculate mean bond enthalpies*
- *perform simple calculations using enthalpy values*

Standard enthalpy of formation, $\Delta H_f^{\ominus}$

The enthalpy change when one mole of a compound is formed from its elements in their standard states. Note that this value is equal to zero for an element.

Hess's law

If a reaction can occur by more than one route, the overall enthalpy change is the same whichever route is taken.

Standard enthalpy changes

Enthalpy change, $\boldsymbol{\Delta H}$, is the heat change measured at constant pressure. The values of enthalpy changes are dependent on the temperature, pressure, and amount of each substance used. We compare enthalpy changes for different reactions by measuring them under standard conditions. Standard enthalpy changes are:

- shown by the symbol $\boldsymbol{\Delta H^{\ominus}}$
- measured in kJ mol^{-1}
- measured at a pressure of 100 kPa
- measured at a temperature of 298 K

Exothermic and endothermic reactions

Reactions are described as **exothermic** or **endothermic** depending on the direction of energy transfer

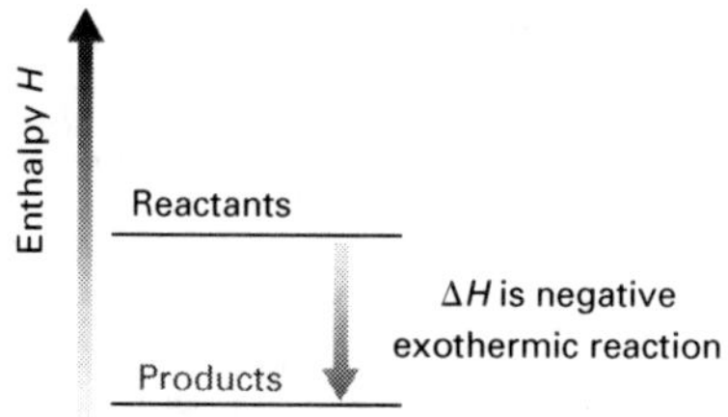

In an exothermic reaction, more energy is released when new bonds form in the products than is needed to break existing bonds in the reactants.

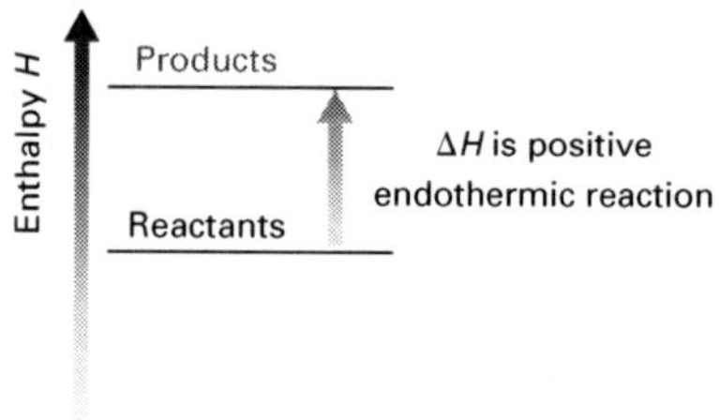

In an endothermic reaction, more energy is needed to break existing bonds in the reactants than is released when new bonds form in the products.

Calculating enthalpy changes

Enthalpy changes can be calculated for reactions by using standard enthalpy changes and Hess's law. Read the data you have been given carefully and use it to construct an enthalpy cycle diagram. Label the arrows and write out your method carefully.

Worked example

Calculate the enthalpy change ($\Delta H_r^{\ominus}$) for the addition of HCl to C_2H_4.

$C_2H_4(g) + HCl(g) \rightarrow C_2H_5Cl(g)$

Given that:

$\Delta H_f^{\ominus}$ $C_2H_4(g)$ = +52 kJ mol^{-1}

$\Delta H_f^{\ominus}$ HCl(g) = –92 kJ mol^{-1}

$\Delta H_f^{\ominus}$ $C_2H_5Cl(g)$ = –105 kJ mol^{-1}

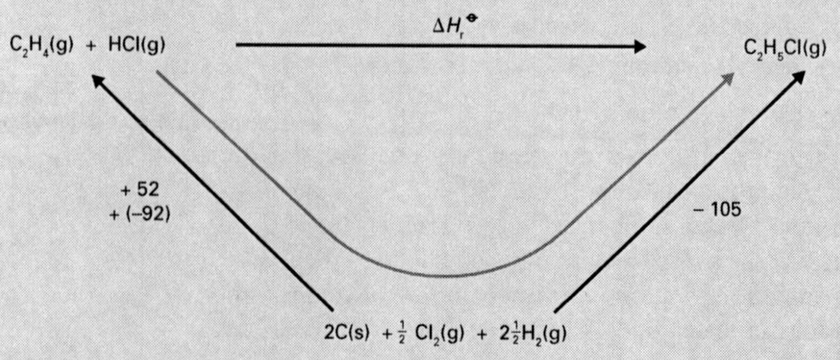

$\Delta H_r^{\ominus} = -[+52+(-92)] + [-105] = -[-40]+[-105] = -65$ kJ mol^{-1}

Note that the sign is reversed when you go against the arrow.

Standard bond dissociation enthalpy, $\Delta H_{diss}^{\ominus}$

When reactions take place bonds are broken and made. The breaking of bonds is an endothermic process (energy is absorbed to break the bonds). The making of bonds is an exothermic process (energy is released when new bonds are formed). Standard bond dissociation enthalpy refers to breaking one mole of bonds under standard conditions producing gaseous fragments:

$Y–Z(g) \rightarrow Y(g) + Z(g)$

For example, the enthalpy change for $HBr(g) \rightarrow H(g) + Br(g)$ is +366 kJ mol^{-1} (i.e. $\Delta H_{diss}^{\ominus}$ HBr = +366 kJ mol^{-1}) so the energy needed to break one mole of H–Br bonds is +366 kJ.

Mean bond enthalpies

Most bonds occur in many substances so the value of the bond enthalpy will vary from one compound to the next. We carry out calculations using mean bond enthalpies, which are the enthalpy change when one mole of a specific bond is broken, averaged over different compounds.

Calculating ΔH from mean bond enthalpy data

ΔH for a reaction can be calculated from mean bond enthalpies by:

- adding together the mean bond enthalpies for the bonds in the reactants
- adding together the mean bond enthalpies for the bonds in the products
- subtracting the second answer from the first

Worked example

Calculate the enthalpy change for the combustion of ethane, $C_2H_6(g)$, using mean bond enthalpy values.

$C_2H_6(g) + 3\frac{1}{2}O_2(g) \rightarrow 2CO_2(g) + 3H_2O(g)$

bond	mean bond enthalpy (kJ mol^{-1})	bond	mean bond enthalpy (kJ mol^{-1})
C–C	348	C–H	413
O=O	498	C=O	743
O–H	463	C=C	612

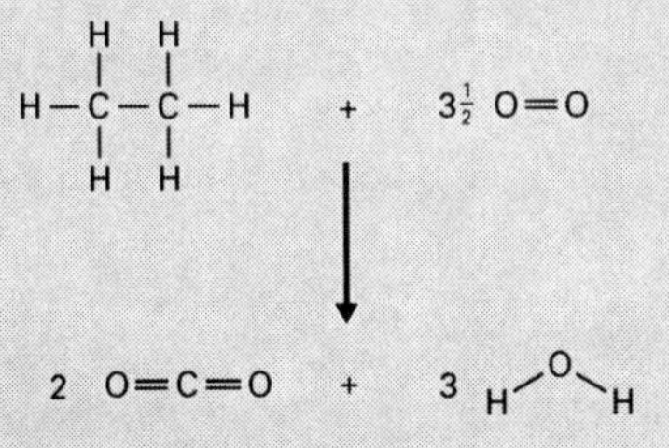

Energy required to break bonds in reactants

1 × C–C = 348 kJ mol^{-1}

6 × C–H = 6 × 413 = 2478 kJ mol^{-1}

$3\frac{1}{2}$ × O=O = $3\frac{1}{2}$ × 498 = 1743 kJ mol^{-1}

Total = 4569 kJ mol^{-1}

Energy released on making bonds in products

4 × C=O = 4 × 743 = 2973 kJ mol^{-1}

6 × O–H = 6 × 463 = 2778 kJ mol^{-1}

Total = 5750 kJ mol^{-1}

Net energy change = 4569 – 5750 = –1181 kJ mol^{-1}

Standard bond dissociation enthalpy, $\Delta H_{diss}^{\ominus}$

The enthalpy change when one mole of bonds of the same type in gaseous molecules is broken under standard conditions, producing gaseous fragments:

$X–Y(g) \rightarrow X(g) + Y(g)$

Bond dissociation enthalpies of the C–H bonds in methane

The bond enthalpy for removal of each C–H bond in methane is slightly different:

$CH_4(g) \rightarrow CH_3(g) + H(g)$
ΔH = 428 kJ mol^{-1}

$CH_3(g) \rightarrow CH_2(g) + H(g)$
ΔH = 470 kJ mol^{-1}

$CH_2(g) \rightarrow CH(g) + H(g)$
ΔH = 416 kJ mol^{-1}

$CH(g) \rightarrow C(g) + H(g)$
ΔH = 338 kJ mol^{-1}

The mean bond enthalpy for C–H in methane is 413 kJ mol^{-1}.

Questions

1 Using the standard enthalpy of formation data below calculate the enthalpy change for the complete combustion of propane:

$C_3H_8(g) + 5O_2(g) \rightarrow 3CO_2(g) + 4H_2O(g)$

$\Delta H_f^{\ominus}$ $C_3H_8(g)$ = –104 kJ mol^{-1}
$\Delta H_f^{\ominus}$ $CO_2(g)$ = –394 kJ mol^{-1}
$\Delta H_f^{\ominus}$ $H_2O(g)$ = –242 kJ mol^{-1}

2 Use the mean bond enthalpies above to calculate the enthalpy change for the cracking of octane:

$C_8H_{18}(l) \rightarrow C_6H_{14}(l) + C_2H_4(g)$

3 Write out each of the definitions on this page five times then cover them up and make sure you can recall them.

5.02 Enthalpy of solution

OBJECTIVES

By the end of this section you should be able to:

- *define lattice enthalpy*
- *define enthalpy of hydration*
- *define enthalpy of solution*
- *calculate enthalpies of solution for ionic compounds*
- *draw enthalpy level diagrams for dissolving*

Lattice enthalpy, $\Delta H_L^\ominus$:

There are two conflicting definitions for this; be careful when answering questions on this topic:

Lattice dissociation enthalpy, $\Delta H_L^\ominus$: the enthalpy change when one mole of an ionic solid is separated into its gaseous ions.

For example:
$KCl(s) \rightarrow K^+(g) + Cl^-(g)$
$\Delta H_L^\ominus = +701$ kJ mol^{-1}

These values are always positive since energy is needed to overcome the strong ionic bonds in the solid.

Lattice formation enthalpy, $\Delta H_L^\ominus$: the enthalpy change when one mole of an ionic solid is formed from its gaseous ions.

For example:
$K^+(g) + Cl^-(g) \rightarrow KCl(s)$
$\Delta H_L^\ominus = -701$ kJ mol^{-1}

These values are always negative.

Enthalpy of hydration, $\Delta H_{hyd}^\ominus$

The enthalpy change when one mole of separate gaseous ions is dissolved completely in water to form one mole of aqueous ions.

For example

$K^+(g) + aq \rightarrow K^+(aq)$

$\Delta H_{hyd}^\ominus = -322$ kJ mol^{-1}

$Cl^-(g) + aq \rightarrow Cl^-(aq)$

$\Delta H_{hyd}^\ominus = -377$ kJ mol^{-1}

When an ionic solid dissolves in water two processes take place.

- The ions in the crystal lattice are separated from each other (an endothermic process).
- Then the separate ions form interactions with the water molecules (exothermic processes).
- As a result, the enthalpy change for dissolving, which is the difference between these two values, may be exothermic or endothermic.

Lattice enthalpy, ΔH_L

Lattice enthalpy may be defined in two different ways. Work through both definitions and make sure you are able to explain them. The values for each process are the same but the signs different.

Lattice dissociation enthalpy is the enthalpy change when one mole of an ionic solid is separated into its gaseous ions; this process is always **endothermic** as energy is absorbed to overcome the strong ionic bonds in the ionic solid.

Lattice formation enthalpy is the enthalpy change when one mole of an ionic solid is formed from its gaseous ions; this process is always **exothermic** as ionic bonds are formed.

Lattice dissociation enthalpies

compound	$\Delta H_L^\ominus$ (kJ mol^{-1})	compound	$\Delta H_L^\ominus$ (kJ mol^{-1})
NaF	918	$MgCl_2$	2493
NaCl	780	MgO	3889
NaBr	742		

Lattice dissociation enthalpy is influenced by the distance between the ions and the charges on the ions.

- The larger the distance between the oppositely charged ions in a crystal lattice the weaker the force of attraction between them.
  As a result the lattice dissociation enthalpy of NaCl is smaller than that of NaF.
- The greater the charge on the ions in a crystal lattice the greater the force of attraction between them.
  As a result the lattice dissociation enthalpy of MgO is much greater than that of $MgCl_2$.

Enthalpy of hydration, $\Delta H_{hyd}^\ominus$

Enthalpy of hydration is the enthalpy change when one mole of separated gaseous ions is dissolved fully in water to form one mole of aqueous ions. The formation of interactions between the ions and water molecules releases energy so this process is exothermic.

Enthalpy of solution, $\Delta H_{soln}^\ominus$

Enthalpy of solution is the enthalpy change when one mole of an ionic substance is dissolved in water so that the ions are totally separated and do not interact with each other. The enthalpy of solution is dependent on both lattice dissociation enthalpy and enthalpies of hydration so may be either endothermic or exothermic.

Calculating enthalpy of solution

When an ionic solid dissolves in water:

- the ionic bonds in the crystal lattice are broken (lattice dissociation enthalpy)
- the separated gaseous ions become surrounded by water molecules (hydration enthalpies)

Enthalpy level diagrams and dissolving

Enthalpy level diagrams can be used to represent many processes. Exothermic processes are shown with arrow heads pointing downwards, endothermic processes with arrows pointing upwards. These can be drawn to scale; the length of each arrow is proportional to the enthalpy change it represents.

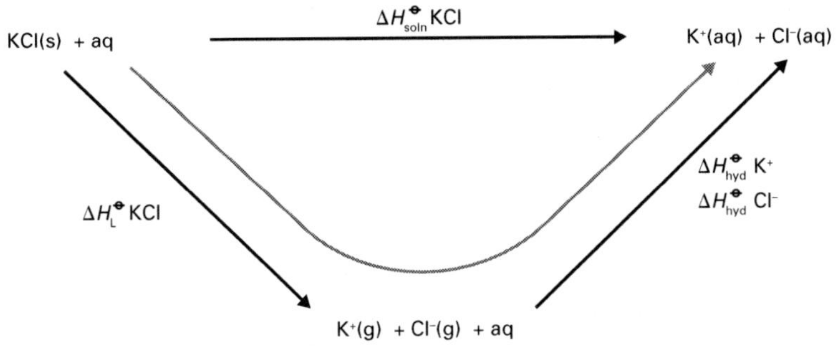

Worked example

Calculate the enthalpy of solution of KCl using the given data

$\Delta H_L^{\ominus}$ KCl = 701 kJ mol^{-1} $\Delta H_{hyd}^{\ominus}$ K^+ = –322 kJ mol^{-1}

$\Delta H_{hyd}^{\ominus}$ Cl^- = –364 kJ mol^{-1}

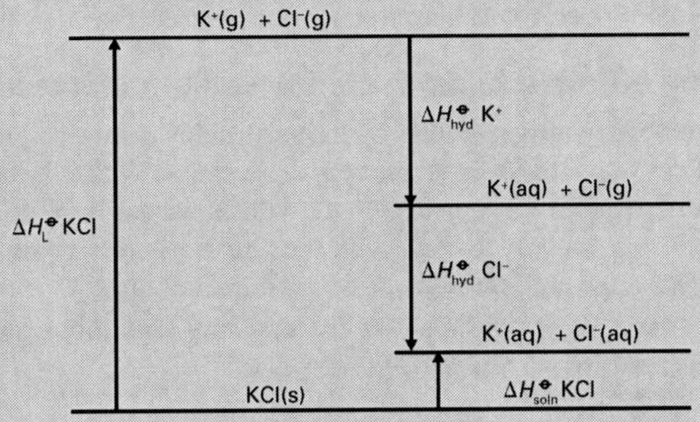

$\Delta H_{soln}^{\ominus}$ KCl = $\Delta H_L^{\ominus}$ KCl + $\Delta H_{hyd}^{\ominus}$ K^+ + $\Delta H_{hyd}^{\ominus}$ Cl^-

$\Delta H_{soln}^{\ominus}$ KCl = +701 + (–322) + (–364) = +15 kJ mol^{-1}

Note that this value is endothermic so the solution becomes colder as the potassium chloride dissolves.

Questions

1 a Draw an enthalpy cycle for the dissolving of ammonium chloride in water.

b Use the data in the table to calculate the enthalpy of solution of ammonium chloride

quantity	value (kJ mol^{-1})
$\Delta H_L^{\ominus}$ NH_4Cl	+640
$\Delta H_{hyd}^{\ominus}$ NH_4^+	–301
$\Delta H_{hyd}^{\ominus}$ Cl^-	–364

c Would ammonium chloride be suitable for use in an instant cold pack? Explain your answer

2 Explain the difference in the following lattice dissociation enthalpies:

compound	value (kJ mol^{-1})
$MgCl_2$	2493
$CaCl_2$	2237
$SrCl_2$	2112

HSW: Cold packs

Solids with endothermic enthalpies of solution are used to make instant cold packs.

The dissolving process

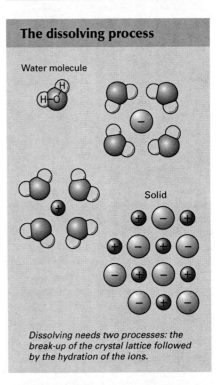

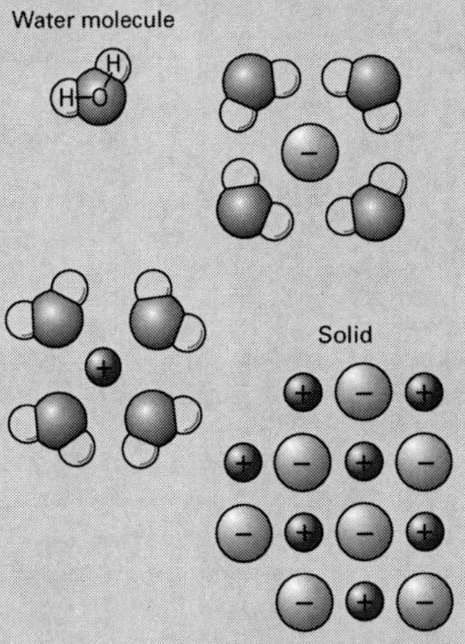

Dissolving needs two processes: the break-up of the crystal lattice followed by the hydration of the ions.

Enthalpy of solution, $\Delta H_{soln}^{\ominus}$

The enthalpy change when one mole of an ionic substance is dissolved in a volume of water large enough to ensure that the ions are separated and do not interact with each other.

For example

$KCl(s) + aq \rightarrow K^+(aq) + Cl^-(aq)$

$\Delta H_{soln}^{\ominus}$ = +17.2 kJ mol^{-1}

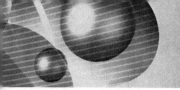

5.03 Born–Haber cycles

OBJECTIVES

By the end of this section you should be able to:

- *define ionization enthalpy*
- *define electron affinity*
- *define enthalpy of atomization*
- *construct and use Born–Haber cycles*
- *compare theoretical and actual lattice enthalpies*

Ionization enthalpies

First ionization enthalpy, $\Delta H_{i1}^{\ominus}$: the enthalpy change when one mole of electrons are removed from one mole of gaseous atoms forming one mole of gaseous 1+ ions.
e.g. $Na(g) \rightarrow Na^+(g) + e^-$

Second ionization enthalpy, $\Delta H_{i2}^{\ominus}$: the enthalpy change when one mole of electrons is removed from one mole of gaseous 1+ ions forming one mole of gaseous 2+ ions.
e.g. $Mg^+(g) \rightarrow Mg^{2+}(g) + e^-$

Electron affinity

First electron affinity, $\Delta H_{ea1}^{\ominus}$: the enthalpy change when one mole of electrons is gained by one mole of gaseous atoms to form one mole of gaseous 1− ions.
e.g. $Cl(g) + e^- \rightarrow Cl^-(g)$

Second electron affinity, $\Delta H_{ea2}^{\ominus}$: the enthalpy change when one mole of electrons is gained by one mole of gaseous 1− ions forming one mole of gaseous 2− ions.
e.g. $O^-(g) + e^- \rightarrow O^{2-}(g)$

Enthalpy of atomization, $\Delta H_{at}^{\ominus}$

The enthalpy change when one mole of gaseous atoms is formed from an element or compound.
e.g. $\frac{1}{2}Cl_2(g) \rightarrow Cl(g)$

$\Delta H_{at}^{\ominus} = +\ 122\ \text{kJ mol}^{-1}$

Note that this value is half that of the bond dissociation enthalpy of chlorine.

A **Born–Haber cycle** is an enthalpy level diagram that enables you to calculate the enthalpy changes involved in the formation of **ionic compounds**. It is essential that you can recall definitions for and use the following enthalpy changes (examples are given for sodium, chlorine, and sodium chloride):

- Enthalpy of formation $Na(s) + \frac{1}{2}Cl_2(g) \rightarrow NaCl(s)$
- Lattice enthalpy $NaCl(s) \rightarrow Na^+(g) + Cl^-(g)$
  Note: Lattice dissociation
- Ionization enthalpy $Na(g) \rightarrow Na^+(g) + e^-$
- Electron affinity $Cl(g) + e^- \rightarrow Cl^-(g)$
- Enthalpy of atomization $\frac{1}{2}Cl_2(g) \rightarrow Cl(g)$

You must be able to use these definitions to draw Born–Haber cycles and then use the cycle to calculate lattice enthalpies.

HSW: Born and Haber

Born–Haber cycles are named after Max Born and Fritz Haber who published their ideas in 1919. Born started his working life in Germany but settled in Britain and shared the Nobel Prize in Physics in 1954 for research into quantum mechanics. Haber won the Nobel Prize in Chemistry in 1918 for developing the Haber Process.

Constructing a Born–Haber cycle for sodium chloride

Born–Haber cycles are always drawn with the enthalpy changes in the same order. Remember that endothermic processes are shown with upwards pointing arrows and exothermic processes are shown with downwards pointing arrows. Cycles can be drawn roughly to scale with the length of each arrow proportional to the enthalpy change but you will not be examined on this.

Work through the enthalpy cycle below making sure you understand each label, and then practise writing it from memory.

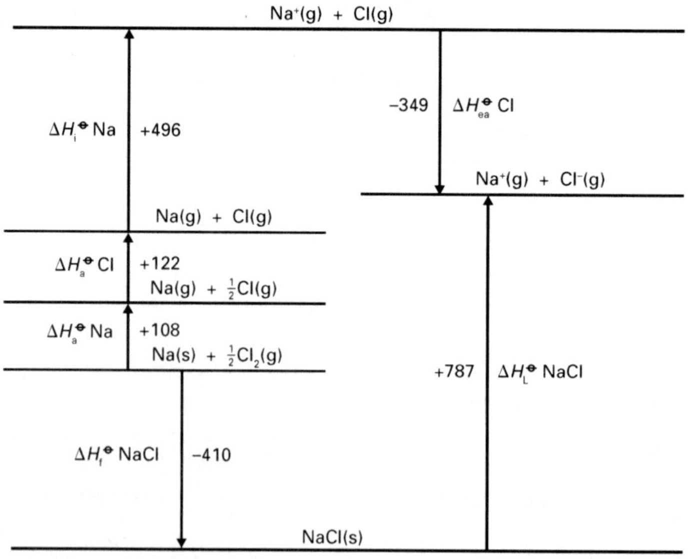

Try covering up one of the values on the cycle then work it out from the rest of the data. Remember to subtract the data if you are moving against the arrow head.

For example: To calculate $\Delta H_L^{\ominus}$ NaCl = −(−410) + 108 + 122 + 496 + (−349) = +787 kJ mol^{-1}

The Born–Haber cycle for calcium chloride

The table below contains the data needed to construct the Born–Haber cycle for calcium chloride, $CaCl_2$. Since this contains Ca^{2+} ions both the first and second ionization enthalpies of calcium need to be used.

standard enthalpy change		value (kJ mol^{-1})
$\Delta H_f^{\ominus}$	Enthalpy of formation of calcium chloride	–795
$\Delta H_{at}^{\ominus}$	Enthalpy of atomization of calcium	193
$\Delta H_{diss}^{\ominus}$	Bond dissociation enthalpy of chlorine	244
$\Delta H_{ie1}^{\ominus}$	First ionization enthalpy of calcium	590
$\Delta H_{ie2}^{\ominus}$	Second ionization enthalpy of calcium	1150
$\Delta H_{ea}^{\ominus}$	First electron affinity of chlorine	–349

Note that the bond dissociation enthalpy of chlorine has been given – this will form two moles of gaseous chlorine atoms so is the same as doubling the enthalpy of atomization.

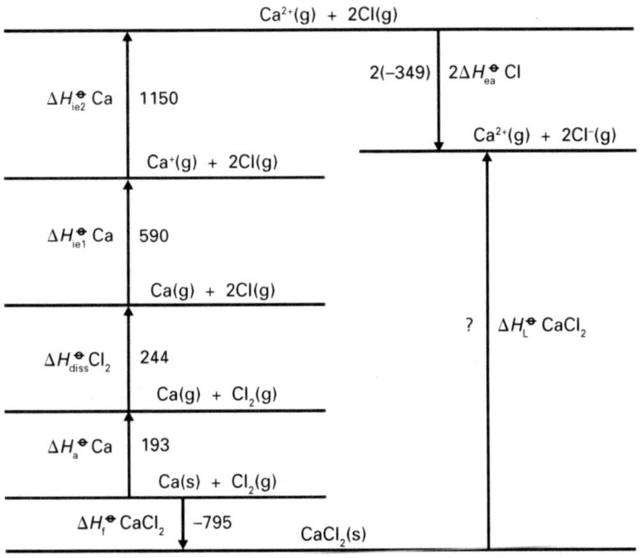

$\Delta H_L^{\ominus}$ $CaCl_2(s)$ = –(–795) + 193 + 244 + 590 + 1150 + 2(–349) = +2274 kJ mo^{l-1}

This value is large so $CaCl_2$ is stable. The lattice dissociation enthalpy for CaCl is +719 kJ mol^{-1} so is less stable.

Covalent character in ionic bonds

The lattice dissociation enthalpies obtained from Born–Haber cycles are described as experimental values since they are calculated from experimental data. Lattice enthalpies can also be calculated from a theoretical model based on the interaction between the ions in the ionic bond.

These interactions include

- the charge on the ions
- the distance from the centre of one ion to the next

Comparison of lattice enthalpies obtained by these two methods show some disagreement.

compound	experimental (Born–Haber) $\Delta H_L^{\ominus}$ (kJ mol^{-1})	theoretical $\Delta H_L^{\ominus}$ (kJ mol^{-1})
AgCl	915	864
AgBr	904	830
AgI	889	808

If the two values are very close then there is evidence in support of the ionic bonding model. If the values differ significantly then there is some distortion of the electron density around the ions introducing a degree of covalent character into the bonding.

The degree of covalent character is described by Fajans' rules. An increased covalent character is caused by:

- small highly charged cations
- large highly charged anions

In the data above AgI has the most covalent character as the iodide ion is largest.

Questions

1 a Draw a Born–Haber cycle and use it to calculate the lattice dissociation enthalpy of MgCl.

enthalpy change	value (kJ mol^{-1})	enthalpy change	value (kJ mol^{-1})
$\Delta H_f^{\ominus}$ MgCl	–94	$\Delta H_{at}^{\ominus}$ Mg	+148
$\Delta H_{diss}^{\ominus}$ Cl_2	+244	$\Delta H_{ie1}^{\ominus}$ Mg	+738
$\Delta H_{ea}^{\ominus}$ Cl	–349	$\Delta H_{ie2}^{\ominus}$ Mg	+1451

b The lattice dissociation enthalpy of $MgCl_2$ is +2524 kJ mol^{-1}. Explain why $MgCl_2$ is formed in preference to MgCl.

2 Use the data given to construct a Born–Haber cycle for sodium oxide, then use your cycle to calculate the second electron affinity of oxygen.

enthalpy change	value (kJ mol^{-1})	enthalpy change	value (kJ mol^{-1})
$\Delta H_f^{\ominus}$ Na_2O	–416	$\Delta H_{at}^{\ominus}$ Na	+109
$\Delta H_{diss}^{\ominus}$ O_2	+496	$\Delta H_{ie1}^{\ominus}$ Na	+495
$\Delta H_{ea1}^{\ominus}$ O	–142	$\Delta H_L^{\ominus}$ Na_2O	–1917

5.04 Free energy and entropy

OBJECTIVES

By the end of this section you should be able to:

- *explain the energy changes when a substance changes state*
- *explain the terms entropy and entropy change*
- *calculate entropy changes from absolute entropy data*
- *explain the meaning of the terms feasibility and free energy*
- *calculate ΔG*
- *use ΔG values to predict if a reaction is feasible*

Entropy diagram for heating water

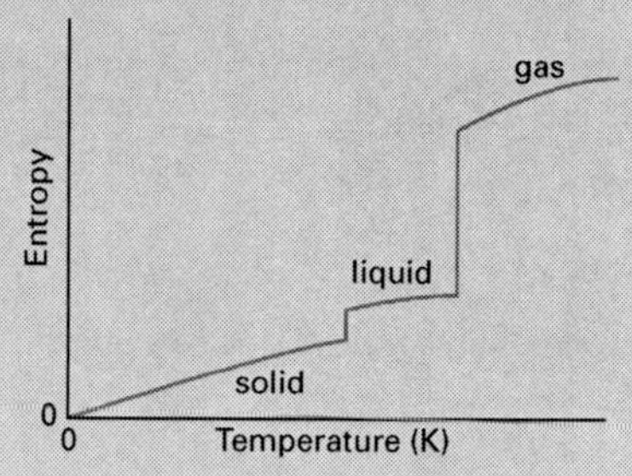

Entropy data for water

Substance	Standard entropy, S (J K^{-1} mol^{-1})
$H_2O(s)$	62.1
$H_2O(l)$	69.9
$H_2O(g)$	188.7

Entropy is a measure of the disorder in a system. An organized room has low **entropy** while a typical teenager's bedroom has high entropy! At absolute zero the entropy of a pure substance is zero.

Entropy

When a substance changes state there is a change in entropy:

- Solids have a regular arrangement of particles.
- The particles in a solid are close to each other.
  As a result solids have low entropies because they are very ordered.
- Liquids have a greater entropy than solids because the particles are disordered – there are many ways they can be arranged.
- Gases have a high entropy because they are highly disordered – the particles can be arranged in a large number of ways.

Examine the diagram showing the entropy changes on heating water. Note the increases in entropy when the state changes. This increase is much larger when the water boils as all the hydrogen bonds between the water particles are broken.

Entropy changes

There will be an overall increase in entropy in a reaction if:

- there are more particles of products than reactants and they are in the same state
- the reactants are solids or liquids and the products include a gas

There is an increase in entropy when the following reactions take place. Work through each one applying the rules above:

$$CaCO_3(s) \rightarrow CaO(s) + CO_2(g)$$
$$KHCO_3(s) + HCl(aq) \rightarrow KCl(aq) + H_2O(l) + CO_2(g)$$
$$C_3H_8(g) + 5O_2(g) \rightarrow 3CO_2(g) + 4H_2O(g)$$

When substances are dissolved in water there is an increase in entropy because the separated ions or molecules in solution have a high degree of disorder.

$CuSO_4(s) + aq \rightarrow Cu^{2+}(aq) + SO_4^{2-}(aq)$

The standard entropy change for a reaction is readily calculated from absolute entropy values using the relationship $\mathbf{\Delta S^{\ominus} = \Sigma S^{\ominus}}$ **(products)** – $\mathbf{\Sigma S^{\ominus}}$ **(reactants)**

Worked example

Calculate the entropy change for the combustion of ethane $C_2H_6(g)$:

$C_2H_6(g) + 3\frac{1}{2}O_2(g) \rightarrow 2CO_2(g) + 3H_2O(g)$

substance	$C_2H_6(g)$	$O_2(g)$	$CO_2(g)$	$H_2O(g)$
entropy, $S^{\ominus}$ (J K^{-1} mol^{-1})	229.5	102.4	213.6	188.7

$\Delta S = \Sigma S^{\ominus}$ (products) – $\Sigma S^{\ominus}$ (reactants)

$\Sigma S^{\ominus}$ (products) = (2 × 213.6) + (3 × 188.7) = 993.3 J K^{-1} mol^{-1}

$\Delta S^{\ominus}$ (reactants) = (229.5) + (3.5 × 102.4) = 587.9 J K^{-1} mol^{-1}

$\Delta S^{\ominus}$ = 993.3 – 587.9 = +405.4 J K^{-1} mol^{-1}

Notice that this is a positive entropy change, which would be expected as the number of molecules has increased.

Feasibility, enthalpy, and entropy

Spontaneous changes are those that happen in only one direction and cannot be reversed without an input of energy:

- Rusting and the reactions of acids with alkalis are spontaneous chemical reactions.

- The dissolving of sugar is a spontaneous process.
- An understanding of enthalpy changes alone is not enough to predict if a process will be spontaneous. We must also consider the entropy changes that occur.

The enthalpy change for a reaction is linked to entropy change by the relationship $\Delta H = T\Delta S$.

For a reaction to occur both the entropy and enthalpy changes must be favourable.

Consider the combustion of ethane under standard conditions:

- The entropy change is positive.
- The enthalpy change is negative.

As a result the reaction is feasible.

Note that the combustion of ethane is feasible but not spontaneous as it has a high activation energy. Ethane will not burn in air until a source of ignition is provided.

Free energy and feasibility

The table below can be used to predict if a reaction is feasible under standard conditions. We often carry reactions out at other temperatures and can determine if a reaction will be feasible by using the idea of **free energy change**.

Feasibility

	ΔS	
ΔH	negative	positive
negative	feasible if $\Delta H < T\Delta S$	always feasible
positive	never feasible	feasible if $\Delta H < T\Delta S$

Free energy change:

- is given the symbol ΔG
- is also called Gibbs energy (after the scientist Willard Gibbs)
- is measured in kJ mol^{-1}
- is calculated by $\Delta G = \Delta H - T\Delta S$

A reaction is feasible when $\Delta G \leq 0$

Worked example

Will the thermal dissociation of NH_4Cl proceed spontaneously at 245 K?

$NH_4Cl(s) \rightarrow NH_3(g) + HCl(g)$ $\Delta H^{\ominus} = +176$ kJmol^{-1} $\Delta S^{\ominus} = +284$ J K^{-1} mol^{-1}

$\Delta G^{\ominus} = \Delta H^{\ominus} - T\Delta S^{\ominus}$

$\Delta G^{\ominus} = 176 - (245 \times 284/1000) = +106$ kJmol^{-1}

$\Delta G^{\ominus}$ is positive so the reaction will not occur spontaneously

Note that $\Delta S^{\ominus}$ is in J K^{-1} mol^{-1} so is converted to kJ mol^{-1}

Entropy units

Entropy and entropy change are measured in J K^{-1} mol^{-1}. Take care using the data as enthalpy and enthalpy changes are measured in kJ mol^{-1}

Worked example

Nitrogen gas can react with oxygen gas to form a mixture of oxides. The temperature at which this reaction will occur is very high. What is the minimum temperature at which nitrogen gas will react spontaneously with oxygen gas?

$\Delta H^{\ominus} = +180$ kJ mol^{-1}

$\Delta S^{\ominus} = +25$ J K^{-1} mol^{-1}

For the reaction to be spontaneous $\Delta G^{\ominus} \leq 0$

Calculate the minimum temperature based on $\Delta G^{\ominus} = 0$

$0 = 180 - T(25/1000)$

$180 = 0.025T$

$T = 180/0.025 = 7200$ K

The minimum temperature at which this reaction will occur is 7200 K

Questions

1. Sketch a labelled diagram showing the entropy changes as solid chlorine is heated until it fully vaporizes.
2. Explain why the reaction of sodium hydrogencarbonate with nitric acid occurs at room temperature even though the reaction is endothermic.
3. Calculate the minimum temperature at which calcium carbonate will decompose to form calcium oxide and carbon dioxide.

   $\Delta H^{\ominus} = +178$ kJ mol^{-1} $\Delta S^{\ominus} = +165$ J K^{-1} mol^{-1}

5.05 Period 3 elements and oxides

OBJECTIVES

By the end of this section you should:

- *be able to describe the trends in atomic radius, ionization enthalpy, melting points, and boiling points of the elements Na–Ar*
- *know the reactions of Na, Mg, Al, Si, P, and S with oxygen*
- *know the reactions of Na–S with water*
- *describe the trends in the reactions of Na and Mg with water*
- *know the links between the physical properties of the oxides of Na–S and their bonding*
- *be able to write equations for reactions of period 3 oxides with simple acids and bases*

Period 3 contains the elements from sodium to argon. Sodium and magnesium are **s block** elements, the other six elements are in the **p block**.

Properties of the period 3 elements

The structure and bonding of the period 3 elements change across the period.

Some key facts are:

- The elements are solids except for chlorine and argon.
- Chlorine and argon are gases at room temperature.
- Sodium, magnesium, and aluminium are metals.
- Silicon is a **metalloid**.
- Phosphorus, sulfur, chlorine, and argon are non-metals.
- The first ionization enthalpy of the elements generally increases across the period.
- The atomic radius decreases across the period.

Make sure you can recall the information in the table below. You learnt about these properties at AS level.

Element	Na	Mg	Al	Si	P_4	S_8	Cl_2	Ar
structure	metallic			giant covalent	simple molecular			monoatomic
bonds broken on melting	metallic			covalent	van der Waals' forces			

Reactions of period 3 elements with water

You must be able to write the equations below showing reactions with water. Note that aluminium and silicon are protected from reacting by a thin layer of their oxides. Each of these elements forms a solution of different pH.

Chlorine: $Cl_2(aq) + H_2O(l) \rightleftharpoons HCl(aq) + HClO(aq)$

A mixture of hydrochloric acid and chloric(I) acid is formed, this mixture is called chlorine water. Chloric(I) acid is a disinfectant used to treat drinking water.

Sodium: $2Na(s) + 2H_2O(l) \rightarrow 2NaOH(aq) + H_2(g)$

Sodium is less dense than water so it floats. It reacts with the water vigorously in an exothermic reaction which releases enough energy to form molten sodium. The solution formed has a pH of 12–14

Magnesium: $Mg(s) + 2H_2O(l) \rightarrow Mg(OH)_2(aq) + H_2(g)$

Magnesium is more dense than water and sinks. It reacts very slowly with water forming a solution of pH 8–10. The magnesium hydroxide is only slightly soluble.

However, magnesium does react vigorously with steam to form magnesium oxide

$Mg(s) + H_2O(g) \rightarrow MgO(s) + H_2(g)$

Reactions of period 3 elements with oxygen

Chlorine and argon do not reaction with oxygen. The other period 3 elements do react and you need to be able to describe their reactions and write equations for them.

Sodium: $4Na(s) + O_2(g) \rightarrow 2Na_2O(s)$

Sodium burns vigorously with a yellow flame forming white sodium oxide.

Magnesium: $2Mg(s) + O_2(g) \rightarrow 2MgO(s)$

Magnesium burns very vigorously with a white flame forming white magnesium oxide.

Aluminium: $4Al(s) + 3O_2(g) \rightarrow 2Al_2O_3(s)$

Aluminium reacts with oxygen forming a thin layer of aluminium oxide, which protects the metal from further attack by oxygen.

Phosphorus: $P_4(s) + 5O_2(g) \rightarrow P_4O_{10}(s)$

Phosphorus exists as different allotropes. Red phosphorus is used in matches but white phosphorus is very hazardous and ignites spontaneously in air, burning with a bright white light and producing white phosphorus(V) oxide.

Sulfur: $S(s) + O_2(g) \rightarrow SO_2(g)$

Sulfur melts readily when heated and burns easily in air with a blue flame forming white fumes of sulfur(IV) oxide.

Silicon reacts with oxygen at 900 °C forming silicon(IV) oxide: $Si(s) + O_2(g) \rightarrow SiO_2(s)$

Properties of the period 3 oxides

Physical properties

The melting points of the period 3 oxides show an increase between sodium oxide and magnesium oxide then a decrease across the rest of the period. This trend is linked to the type of bonding within each oxide.

Note the following trends in properties as we move across period 3:

- The oxides formed change from giant ionic structures to molecular covalent structures.
- Silicon oxide has a giant covalent structure.
- The empirical formula of silicon oxide is SiO_2.
- The oxides change from basic to acidic.
- Aluminium ions are very small and highly charged so polarize oxide ions.

  As a result aluminium oxide has significant covalent character.
- So aluminium oxide is **amphoteric** (it has both acidic and basic properties).

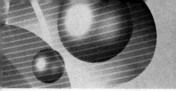

oxide	Na_2O	MgO	Al_2O_3	SiO_2	P_4O_{10}	SO_2	SO_3
typical pH in water	14	10	7	7	0	3	0
structure	giant				molecular		
bonding	ionic		ionic with covalent character	covalent			
nature of oxide	basic		amphoteric	acidic			

The trend of the formulae of period 3 oxides can be explained by considering the highest oxidation states of the period 3 elements. This highest oxidation state increases from +1 for sodium (which is in group 1) through to +7 for chlorine (which is in group 7).

This means that the highest oxides of the elements from sodium to sulfur are: Na_2O, MgO, Al_2O_3, SiO_2, P_4O_{10}, and SO_3.

Note that there are two possible oxides of sulfur: sulfur(IV) oxide, SO_2, and sulfur(VI) oxide, SO_3.

As expected, the ionic substances only conduct electricity when molten. Magnesium oxide and aluminium oxide do not conduct electricity when added to water owing to their low degree of solubility.

Reactions with water

Practise writing each of these equations and make sure you can recall the pH of the solutions formed.

$Na_2O(s) + H_2O(l) \rightarrow 2NaOH(aq)$ pH = 14

$MgO(s) + H_2O(l) \rightarrow Mg(OH)_2(aq)$ pH = 10

MgO sparingly soluble

Aluminium oxide and silica are insoluble in water and do no react with it.

$P_4O_{10}(s) + 6H_2O(l) \rightarrow 4H_3PO_4(aq)$ pH = 0

Vigorous reaction

$SO_2(g) + H_2O(l) \rightleftharpoons H_2SO_3(aq)$ pH = 1

Reacts readily with H_2O

$SO_3(g) + H_2O(l) \rightarrow H_2SO_4(aq)$ pH = 0

Reacts vigorously

Sodium oxide and magnesium oxide are basic; they react with acids forming salts and water:

$Na_2O(s) + H_2SO_4(aq) \rightarrow Na_2SO_4(aq) + H_2O(l)$

$MgO(s) + 2HCl(aq) \rightarrow MgCl_2(aq) + H_2O(l)$

Remember aluminium oxide is amphoteric.

$Al_2O_3(s) + 3H_2SO_4(aq) \rightarrow Al_2(SO_4)_3(s) + 3H_2O(l)$

$Al_2O_3(s) + 2NaOH(aq) + 3H_2O(l) \rightarrow 2NaAl(OH)_4(aq)$

Silica is insoluble in water but does react with bases:

$SiO_2(s) + 2NaOH(aq) \rightarrow Na_2SiO_3(aq) + H_2O(l)$

Phosphorus(V) oxide and sulfur(IV) oxide react with bases to form salts.

For example:

$P_4O_{10}(s) + 12NaOH(aq) \rightarrow 4Na_3PO_4(aq) + 6H_2O(l)$

$SO_2(g) + CaCO_3(s) \rightarrow CaSO_3(s) + CO_2(g)$

Melting and boiling points of period 3 elements

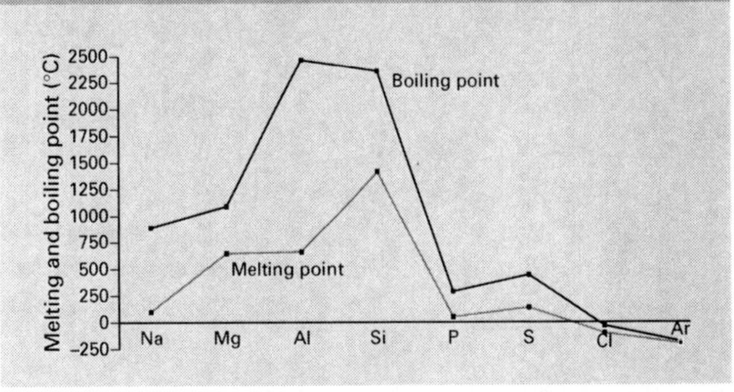

First ionization enthalpy across period 3

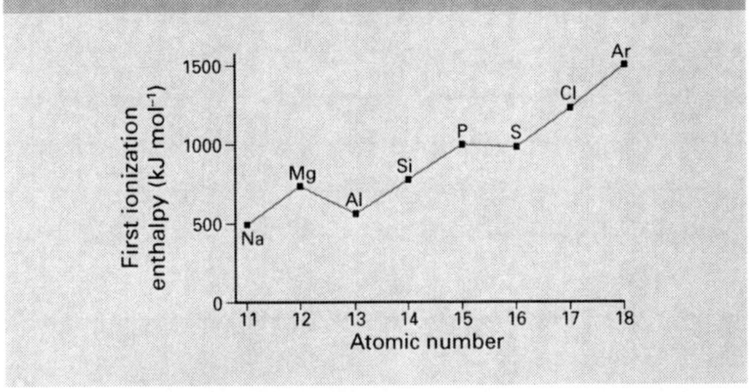

Melting points of period 3 oxides

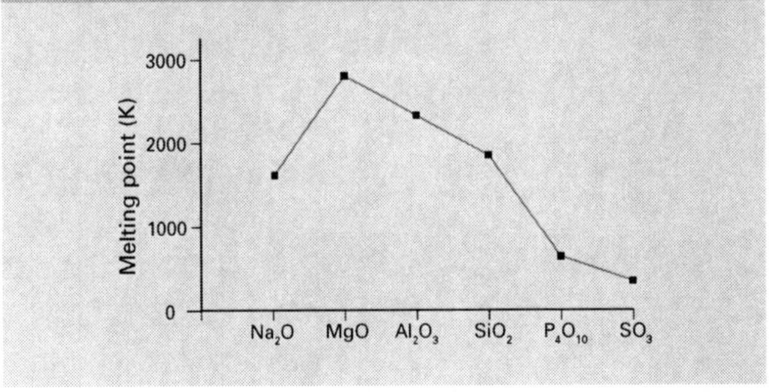

Questions

1. Write balanced equations including state symbols for the following reactions:
   - **a** magnesium oxide and sulfuric acid
   - **b** sulfur(IV) oxide and sodium carbonate
   - **c** aluminium oxide and sulfuric acid
2. Using the graph given explain the trends in boiling point across period 3.
3. Explain, using equations in your answer, the reactions between period 3 elements and water.

5.06 Redox reactions

OBJECTIVES

By the end of this section you should:

- *know that oxidation is loss of electrons and reduction is gain of electrons*
- *be apply to work out the oxidation state of an element in a compound or ion*
- *be able to write and balance half-equations*
- *be able to combine simple half-equations to give overall equations*

Redox reactions involve both the loss of electrons (oxidation) and the gain of electrons (reduction). These processes always occur at the same time.

Oxidation states

Oxidation states are used to find out the number of electrons that must be gained to make a neutral atom. For example, the oxidation state of magnesium in Mg^{2+} is +2 because two electrons would need to be gained in order to make a neutral atom:

$$Mg^{2+} + 2e^- \rightarrow Mg$$

The oxidation state of chlorine in Cl^- is –1 because one electron would need to be lost to make a neutral atom:

$$Cl^- \rightarrow Cl + e^-$$

It is important that you can apply the oxidation number rules. Work through the table below and make sure you can remember each one then work through the examples below to determine the oxidation states of various elements.

Element	Oxidation state in compounds and ions	Exceptions
H	+1	Metal hydrides: oxidation state is –1, e.g. NaH
Li, Na, K	+1	
Mg, Ca, Ba	+2	
Al	+3	
F	–1	
Cl	–1	Oxidation state varies when combined with F or O
O	–2	Oxidation state is not –2 when combined with F Oxidation state is –1 in peroxides e.g. H_2O_2

Step 1 Write down the formula.

Step 2 Using the table write the known oxidation states above the symbol for each element.

Step 3 Using the known oxidation states work out the total oxidation state.

Step 4 Subtract the total oxidation states from zero for compounds or from the charge if the problem is about a polyatomic ion.

Worked example 1	
What is the oxidation state of aluminium in Al_2O_3?	
Step 1	Al_2O_3
Step 2	$Al_2\overset{-2}{O}_3$
Step 3	$Al_2\underset{-6}{\overset{-2}{O}}_3$
Step 4	Total oxidation state of Al = 0 – (–6) = +6 Oxidation state of each Al = +6/2 = +3

Worked example 2	
What is the oxidation state of manganese in MnO_4^-?	
Step 1	MnO_4^-
Step 2	$Mn\overset{-2}{O}_4^-$
Step 3	$Mn\underset{-8}{\overset{-2}{O}}_4^-$
Step 4	Total oxidation state of Mn = –1–(–8) = +7

OIL RIG

Oxidation **i**s the **l**oss of electrons

Oxidation is shown by an increase in oxidation number

Reduction **i**s the **g**ain of electrons

Reduction is shown by a decrease in oxidation number

d block elements

d block elements are those with electrons in the d sub-level.

Transition elements are d block elements that form at least one stable ion with a partially filled d sub-level.

Half-equations

Half-equations show the loss or gain of electrons by one substance. For example, this is the full equation for the reaction between calcium and chlorine. Note that it can be split into two half-equations:

$$Ca(s) + Cl_2(g) \rightarrow CaCl_2(s)$$

The oxidation half-equation is: $Ca \rightarrow Ca^{2+} + 2e^-$

The reduction half-equation is: $Cl_2 + 2e^- \rightarrow 2Cl^-$

At A2 level you need to be able to

- construct and balance simple half-equations
- combine half-equations to show overall reactions

Work through the two examples below showing how to balance half-equations.

Combining half-equations

Half-equations are straightforward to combine as long as you remember that the electrons on the left hand side and the right hand side of the equation must cancel out.

Worked example 1

Complete the following half-equation: $Br_2 \rightarrow Br^-$

Step 1 Write out the two species $Br_2 \rightarrow Br^-$

Step 2 Balance the atoms $Br_2 \rightarrow \mathbf{2}Br^-$

Step 3 Add enough electrons to make the total charges the same on each side
$Br_2 + \mathbf{2e^-} \rightarrow 2Br^-$

Worked example 2

Complete the following half-equation: $Cr_2O_7^{2-} \rightarrow Cr^{3+}$

Half-equations involving d block elements may involve hydrogen ions. These are needed when the d block element is combined with oxygen in a polyatomic ion.

Step 1 Write out the two species $Cr_2O_7^{2-} \rightarrow Cr^{3+}$

Step 2 Add enough water molecules to the right hand side to balance the oxygen atoms $Cr_2O_7^{2-} \rightarrow Cr^{3+} + \mathbf{7H_2O}$

Step 3 Add enough H^+ ions to the left hand side to account for the hydrogen atoms on the right $Cr_2O_7^{2-} + \mathbf{14H^+} \rightarrow Cr^{3+} + 7H_2O$

Step 4 Balance all the atoms in the equation $Cr_2O_7^{2-} + 14H^+ \rightarrow 2Cr^{3+} + 7H_2O$

Step 5 Add enough electrons to make the total charges on each side the same
$Cr_2O_7^{2-} + 14H^+ + \mathbf{6e^-} \rightarrow 2Cr^{3+} + 7H_2O$

Worked example

Combine these two half-equations:

$Cr_2O_7^{2-} + 14H^+ + 6e^- \rightarrow 2Cr^{3+} + 7H_2O$ $Fe^{2+} \rightarrow Fe^{3+} + e^-$

$Cr_2O_7^{2-} + 14H^+ + 6e^- \rightarrow 2Cr^{3+} + 7H_2O$ $\mathbf{6}Fe^{2+} \rightarrow \mathbf{6}Fe^{3+} + \mathbf{6}e^-$

Add together and cancel electrons:

$Cr_2O_7^{2-} + 14H^+ + 6Fe^{2+} \rightarrow 2Cr^{3+} + 6Fe^{3+} + 7H_2O$

Ions of vanadium

- Vanadium has the electron configuration [Ar] $4s^2\ 3d^3$.
- It is able to form ions with a variety of oxidation states.
- Make sure you can work out the oxidation state of vanadium in each one.

ion	oxidation state of vanadium
VO_2^+	+5
VO^{2+}	+4
V^{3+}	+3
V^{2+}	+2

Questions

1 Work out the oxidation states of the element in bold in the following:

a $\mathbf{Mn}O_4^-$

b $\mathbf{Fe}^{2+}$

c $\mathbf{Cl}O_3^-$

d $\mathbf{N}H_4^+$

2 Balance the following half-equations:

a $Cu^{2+} \rightarrow Cu^+$

b $MnO_4^- \rightarrow Mn^{2+}$

c $O^{2-} \rightarrow O_2$

3 Combine the pairs of half-equations below:

a $Mg \rightarrow Mg^{2+} + 2e^-$
$Br_2 + 2e^- \rightarrow 2Br^-$

b $Zn \rightarrow Zn^{2+} + 2e^-$
$Cr_2O_7^{2-} + 14H^+ + 6e^- \rightarrow 2Cr^{3+} + 7H_2O$

5.07 Electrochemical cells

OBJECTIVES

By the end of this section you should:

- *be able to define standard electrode potential (electromotive force)*
- *know that standard electrode potentials are measured relative to the standard hydrogen electrode*
- *be able to draw electrochemical cells*
- *be able to construct cell diagrams for simple electrochemical cells*

When a strip of magnesium is placed into a solution of copper(II) sulfate, a redox reaction occurs:

$$Mg(s) + Cu^{2+}(aq) \rightarrow Mg^{2+}(aq) + Cu(s)$$

The magnesium is oxidized: $Mg(s) \rightarrow Mg^{2+}(aq) + 2e^-$

The copper(II) ions are reduced $Cu^{2+}(aq) + 2e^- \rightarrow Cu(s)$

- The sulfate ions are spectator ions and do not take part in the reaction. As a result they are not included in the equations for the redox reaction.

If instead you were to put copper metal into a solution of magnesium(II) sulfate, no reaction would occur. An understanding of electrochemical cells enables you to predict whether a reaction will occur or not.

Half-cells

The standard electrode potential, $E^{\ominus}$, for a redox reaction tells us whether a substance is readily oxidized or reduced.

To determine the $E^{\ominus}$ value a **half-cell** is made and connected to the standard hydrogen half-cell.

This hydrogen half-cell is assigned an $E^{\ominus}$ value of 0.00 V so every other half-cell is measured relative to hydrogen.

The diagram shows how the $E^{\ominus}$ value for a copper half-cell could be determined

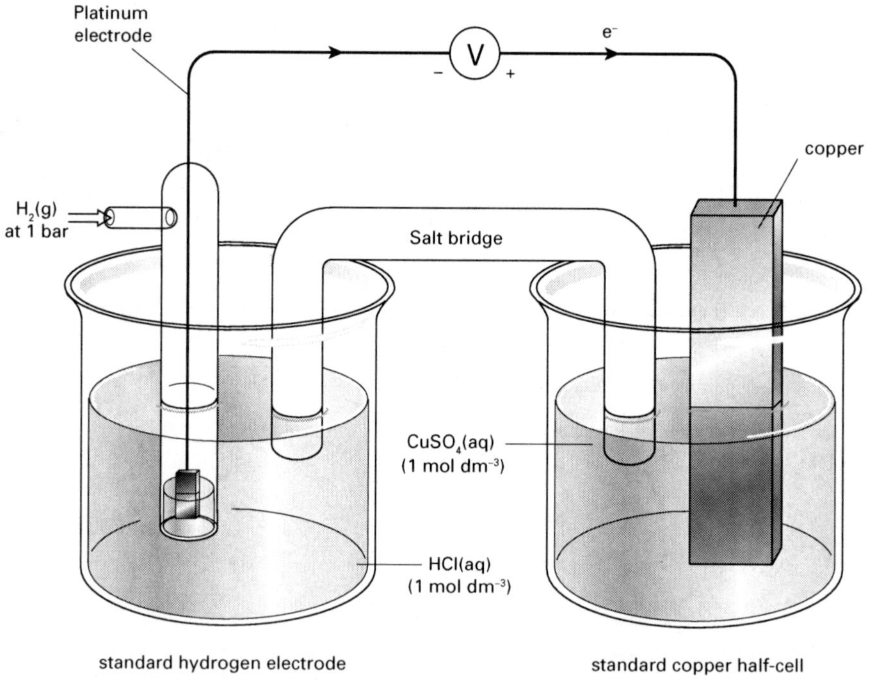

The $E^{\ominus}$ value for the copper half-cell is +0.34 V. The positive sign shows that the potential on the copper electrode is more positive than that on the standard hydrogen electrode. The electrons flow from the hydrogen half-cell to the copper half-cell.

Note that the equations representing half-cells are always written as **reduction reactions** with the use of an equilibrium arrow.

Electrochemical cells

An **electrochemical cell** can be constructed for any pair of half-cells.

The diagram shows the electrochemical cell obtained when the magnesium and copper half-cells are connected together.

Standard hydrogen electrode

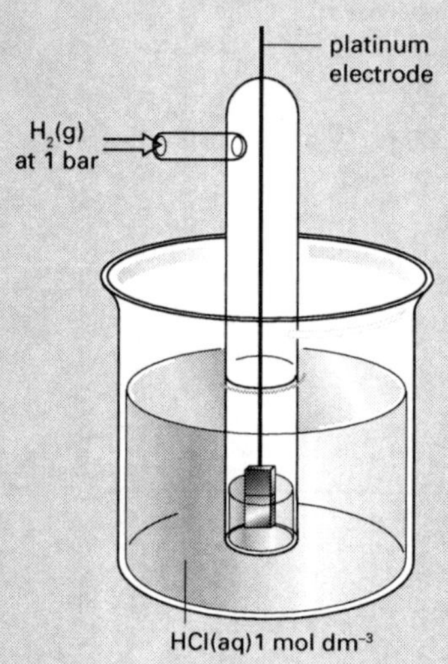

Standard electrode potential

The standard electrode potential, $E^{\ominus}$, is the electromotive force of an electrochemical cell measured relative to the hydrogen half-cell under standard conditions, i.e.:

- a temperature of 298 K (25 °C)
- a pressure of 100 kPa (100,000 Pa)
- solutions of ions at a concentration of 1.00 mol dm^{-3}

The hydrogen half-cell is assigned an $E^{\ominus}$ = 0 V ($2H^+(aq) + 2e^- \rightleftharpoons H_2(g)$)

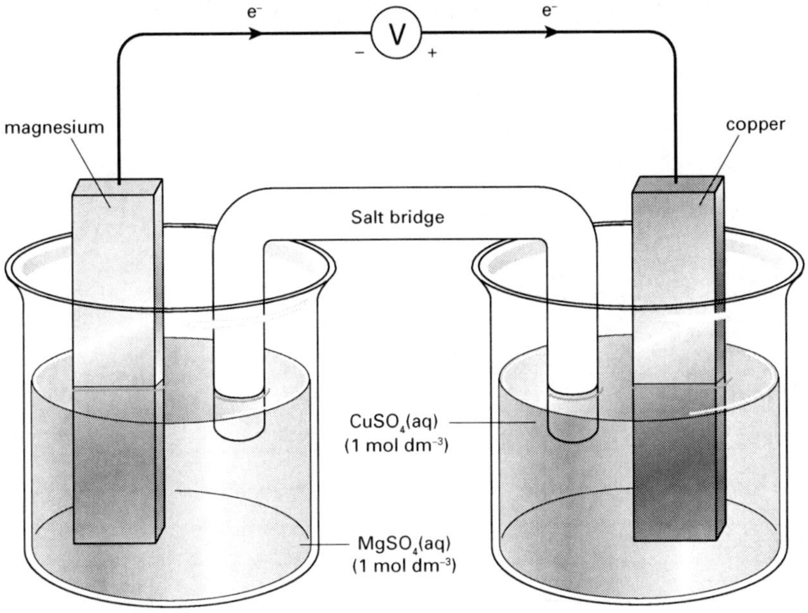

When this cell is connected up:

- the concentration of magnesium ions increases
- solid copper is deposited on the copper electrode
- a reading of +2.71 V is obtained

This tells us that

- In the magnesium half-cell oxidation occurs so the equilibrium $Mg^{2+}(aq) + 2e^- \rightleftharpoons Mg(s)$ lies to the left hand side.
- The $E^{\ominus}$ value for the magnesium half-cell is –2.37 V.
- In the copper half-cell reduction occurs so the equilibrium $Cu^{2+}(aq) + 2e^- \rightleftharpoons Cu(s)$ lies to the right hand side.
- The $E^{\ominus}$ value for the magnesium half-cell is +0.34 V
- Electrons flow from the magnesium half-cell to the copper half-cell.

In the next section you will revise how to calculate and interpret the voltmeter reading.

Cell diagrams

Cell diagrams are used to represent electrochemical cells in a simpler way.

The cell diagram for the experiment above is:

$$Mg(s) \mid Mg^{2+}(aq) \parallel Cu^{2+}(aq) \mid Cu(s)$$

The single line separates the two phases and the double vertical line represents the salt bridge.

If the half-cell contains a mixture of aqueous ions, a comma is used to separate them. For example:

$$Fe(s) \mid Fe^{2+}(aq), Fe^{3+}(aq)$$

Salt bridge

The salt bridge in a simple cell is a piece of filter paper that has been soaked in saturated potassium chloride or potassium nitrate solution. The role of the salt bridge is to balance the charges of the electrochemical cell. Positive ions move into the half-cell in which reduction occurs and negative ions move into the half-cell in which oxidation occurs.

Questions

1 Draw a labelled diagram to show how the standard electrode potential of the zinc half-cell could be determined ($Zn^{2+}(aq) + 2e^- \rightleftharpoons Zn(s)$)

2 Draw a labelled diagram showing the electrochemical cell obtained when the zinc and copper half-cells are connected.

5.08 The electrochemical series

OBJECTIVES

By the end of this section you should:

- *know that standard electrode potentials can be listed as an electrochemical series*
- *be able to identify oxidizing agents and reducing agents using standard electrode potential data*
- *be able to use $E^{\ominus}$ values to calculate the e.m.f of a cell and to predict the direction of simple redox reactions*

Oxidizing ability

The strongest oxidizing agents have the most positive $E^{\ominus}$ values. A species on the left hand side can gain electrons from any of the species below and to the right, and so oxidize them.

Reducing ability

The strongest reducing agents have the most negative $E^{\ominus}$ values. A species on the right hand side can donate electrons to any species above and to the left, and so reduce them.

Standard electrode potentials, which are measured by connecting a half-cell to a standard hydrogen electrode, can be used to calculate the e.m.f. of a cell and to predict the direction of simple redox reactions.

The electrochemical series

electrode reaction	E (V)
$F_2(g) + 2e^- \rightleftharpoons 2F^-(aq)$	+2.87
$MnO_4^-(aq) + 8H^+(aq) + 5e^- \rightleftharpoons Mn^{2+}(aq) + 4H_2O(l)$	+1.51
$Cl_2(aq) + 2e^- \rightleftharpoons 2Cl^-(aq)$	+1.36
$Cr_2O_7^{2-}(aq) + 14H^+(aq) + 6e^- \rightleftharpoons 2Cr^{3+}(aq) + 7H_2O(l)$	+1.33
$Br_2(aq) + e^- \rightleftharpoons 2Br^-(aq)$	+1.09
$Fe^{3+}(aq) + e^- \rightleftharpoons Fe^{2+}(aq)$	+0.77
$I_2(aq) + 2e^- \rightleftharpoons 2I^-(aq)$	+0.54
$Cu^+(aq) + e^- \rightleftharpoons Cu(s)$	+0.52
$Cu^{2+}(aq) + 2e^- \rightleftharpoons Cu(s)$	+0.34
$Cu^{2+}(aq) + e^- \rightleftharpoons Cu^+(aq)$	+0.15
$2H^+(aq) + 2e^- \rightleftharpoons H_2(g)$	0.00
$Pb^{2+}(aq) + 2e^- \rightleftharpoons Pb(s)$	−0.13
$Fe^{2+}(aq) + 2e^- \rightleftharpoons Fe(s)$	−0.44
$Zn^{2+}(aq) + 2e^- \rightleftharpoons Zn(s)$	−0.76
$Mg^{2+}(aq) + 2e^- \rightleftharpoons Mg(s)$	−2.37
$Na^+(aq) + e^- \rightleftharpoons Na(s)$	−2.71
$K^+(aq) + e^- \rightleftharpoons K(s)$	−2.92
$Li^+(aq) + e^- \rightleftharpoons Li(s)$	−3.03

Using the electrochemical series

The list of standard electrode potentials in order is called the electrochemical series. This is written in order of increasingly negative potential or increasingly positive potential.

Examine the electrochemical series on this page carefully. You need to be able to

- combine half-equations to write overall equations
- identify oxidizing agents and reducing agents for pairs of equations

Consider the half-equation and $E^{\ominus}$ value for chlorine/chloride

- The $E^{\ominus}$ value is positive.
- This means that the equilibrium lies a long way to the right hand side. As a result chlorine is readily reduced to chloride ions.
- Consider the half-equation and $E^{\ominus}$ value for zinc ions/zinc.
- The $E^{\ominus}$ value is negative.
- This means that the equilibrium lies a long way to the left hand side. As a result zinc is readily oxidized to zinc ions.

If we want to combine these two half-equations then we must reverse the half-equation for the zinc system before we add the equations together:

Reduction half-equation $Cl_2(aq) + 2e^- \rightarrow 2Cl^-(aq)$

Oxidation half-equation $Zn(s) \rightarrow Zn^{2+}(aq) + 2e^-$

Overall equation $Cl_2(aq) + Zn(s) \rightarrow Zn^{2+}(aq) + 2Cl^-(aq)$

If the number of electrons provided by the oxidation is not equal to the number needed for the reduction then the equations must be balanced carefully so that the number of electrons cancels on either side.

Calculating cell e.m.f.

The e.m.f. of a cell is calculated simply from the cell diagram and the $E^{\ominus}$ values. You will see this calculated in many different ways in different books.

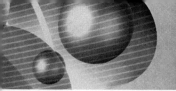

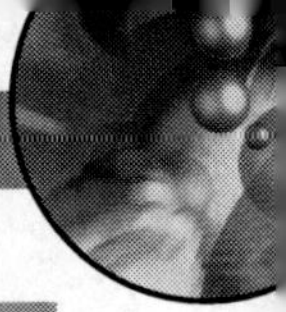

This method matches that used in the A2 Chemistry for AQA student text book. If you have been taught this by another method then refer to the method you have been taught!

Example *Calculate the e.m.f of this cell* Mg(s) | Mg^{2+}(aq) || Cu^{2+}(aq) | Cu(s)

$Mg^{2+}(aq) + 2e^- \rightarrow Mg(s)$ $\quad E^\ominus = -2.37$ V

$Cu^{2+}(aq) + 2e^- \rightarrow Cu(s)$ $\quad E^\ominus = +0.34$ V

$E^\ominus_{cell} = E^\ominus_{right} - E^\ominus_{left} = (+0.34) - (-2.37) = +2.71$ V

Predicting the direction of spontaneous change

The easiest method for working out the direction of spontaneous change is to remember that the spontaneous change is the one which produces a positive $E^\ominus_{cell}$ value. For this to occur, the half-cell with the more positive electrode potential must be on the right hand side.

Work through this example to practise predicting the direction of spontaneous change. The diagram shown represents the magnesium and copper cell.

Step 1 Draw two vertical lines a few centimetres apart: these represent the electrodes.

Step 2 Label the left hand line as negative and the right hand line as positive.

Step 3 Draw a spiral line from the bottom of the left hand electrode to the bottom of the right hand electrode. Add an arrow pointing to the right and label it e^-: this shows the flow of electrons.

Step 4 Write the equation for the left hand electrode next to the left hand electrode, and the equation for the right hand electrode next to the right hand electrode.

Step 5 Add the two $E^\ominus$ values to the two electrodes.

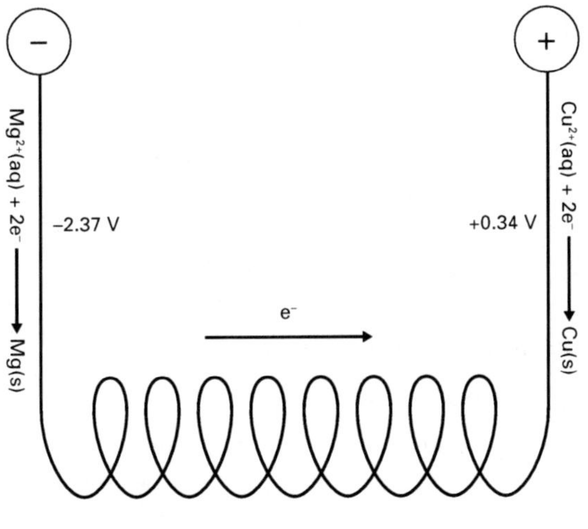

Note that

- In the diagram, electrons flow from left to right.
- Oxidation happens on the left (this is the half-cell with the more negative electrode potential).
- Reduction happens on the right (this is the half-cell with the more negative electrode potential).

Since oxidation occurs at the left hand electrode, the equation for the reaction is reversed.

Left hand electrode reaction: $Mg(s) \rightarrow Mg^{2+}(aq) + 2e^-$

Right hand electrode reaction: $Cu^{2+}(aq) + 2e^- \rightarrow Cu(s)$

Overall cell equation: $Mg(s) + Cu^{2+}(aq) \rightarrow Mg^{2+}(aq) + Cu(s)$

This represents the spontaneous cell reaction.

$E^\ominus_{cell} = E^\ominus_{right} - E^\ominus_{left} = (+0.34) - (-2.37) = +2.71$ V

Large and positive $E^\ominus$ values indicate that a reaction is feasible. However, the reaction may not occur spontaneously as the activation energy may be high.

Cell e.m.f.

The e.m.f. can be determined from a cell diagram and the relevant standard electrode potential data. Use the cell diagram to see which half-cell is on the right and which one is on the left.

$E^\ominus_{cell} = E^\ominus_{right} - E^\ominus_{left}$

Questions

1 For each of the pairs of substances below, state which one is the more powerful reducing agent.

  a Li or Zn

  b F^- or Br^-

2 For each of the following pairs of redox reactions, write the equation for the spontaneous redox reaction, identify the reduced species, and calculate $E^\ominus_{cell}$. Use the data from the electrochemical series to help you.

  a Fe^{2+}/Fe and Zn^{2+}/Zn

  b Br_2/Br^- and I_2/I^-

  c MnO_4^-/Mn^{2+} and Fe^{3+}/Fe^{2+}

5.09 Commercial cells

OBJECTIVES

By the end of this section you should:

- understand that electrochemical cells can be used as a commercial source of electrical energy
- appreciate that cells can be non-rechargeable, rechargeable or fuel cells
- be able to deduce the reaction occurring in non-rechargeable and rechargeable cells and to calculate the cell's e.m.f.
- understand the electrode reactions of a hydrogen–oxygen fuel cell
- appreciate the benefits and risks to society associated with the use of electrochemical cells

A typical zinc–carbon cell

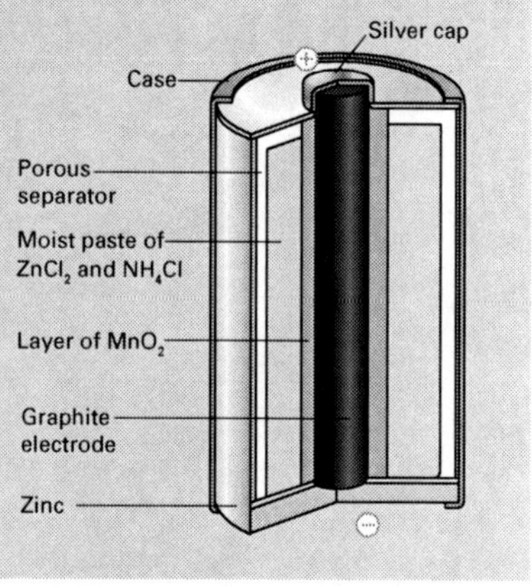

A typical alkaline dry cell

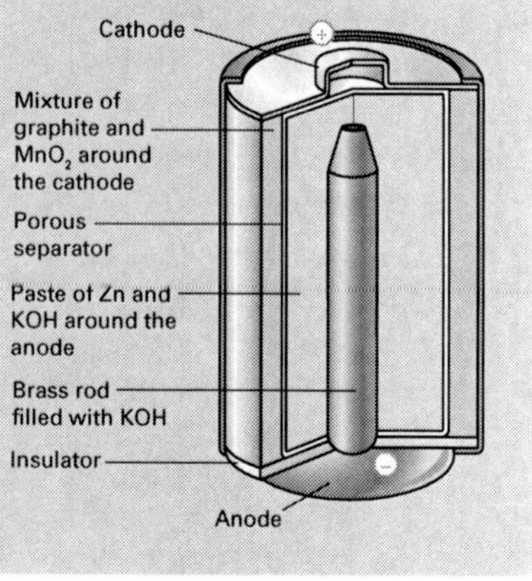

Electrochemical cells are a convenient source of electrical energy. A battery is made from two or more cells connected in series.

There are three main types of commercial electrochemical cell:

- Primary cells – These are not rechargeable and are thrown away when run down.
- Secondary cells – These can be recharged after they run down.
- Fuel cells – These produce electricity from gaseous or liquid fuels.

Early primary cells were wet cells. For example, the Daniell cell invented by John Daniell in 1836 consisted of a piece of zinc dipped into aqueous zinc sulfate in a porous pot surrounded by aqueous copper sulfate in a copper can.

Look back at the electrode potentials on Spread 5.07. Zinc is the **anode** and the copper the **cathode**. Overall $Zn(s) + Cu^{2+}(aq) \rightarrow Zn^{2+}(aq) + Cu(s)$ $E^{\ominus} = +1.10$ V

Most modern cells are dry cells in which the electrolyte is present as a damp paste or gel, which cannot be spilt if the cell is turned upside down.

Zinc–carbon cells

The anode of a zinc–carbon cell is a zinc can, which contains a paste of ammonium chloride and zinc chloride. The cathode is a mixture of powdered manganese(IV) oxide and graphite surrounding a graphite rod.

The reactions which occur in the zinc–carbon cell are:

At the anode $Zn(s) + 2NH_3(aq) \rightarrow [Zn(NH_3)_2]^{2+}(aq) + 2e^-$

At the cathode $2MnO_2(s) + 2H^+(aq) + 2e^- \rightarrow Mn_2O_3(s) + H_2O(l)$

Overall

$Zn(s) + 2MnO_2(s) + 2NH_4^+(aq) \rightarrow [Zn(NH_3)_2]^{2+}(aq) + Mn_2O_3(s) + H_2O(l)$

Alkaline dry cells

Alkaline cells produce the same potential difference as a zinc–carbon cell but they last longer. The electrolyte in the cell is potassium hydroxide. The anode is powdered zinc alloy mixed with potassium hydroxide. The cathode is a mixture of potassium hydroxide, powdered manganese(IV) oxide, and graphite.

At the anode: $Zn(s) + 2OH^-(aq) \rightarrow ZnO(s) + H_2O(l) + 2e^-$

At the cathode $2MnO_2(s) + H_2O(l) + 2e^- \rightarrow Mn_2O_3(s) + 2OH^-(aq)$

Overall $Zn(s) + 2MnO_2(s) \rightarrow ZnO(s) + 2Mn_2O_3(s)$

The lead–acid battery

The lead–acid battery was invented in 1859 by a French scientist, Gaston Plante. A typical car battery has six cells in series, each producing a voltage of 2 V, so the battery has an overall voltage of 12 V.

The electrolyte is 6 mol dm^{-3} H_2SO_4

When charged, the anode is spongy lead covering a grid made of lead alloy; the cathode is lead(IV) oxide also in a grid. Using a grid gives a large surface area.

Anode reactions

Under discharge: $Pb(s) + SO_4^{2-}(aq) \rightarrow PbSO_4(s) + 2e^-$

When charging: $PbSO_4(s) + 2e^- \rightarrow Pb(s) + SO_4^{2-}(aq)$

Cathode reactions

Under discharge: $PbO_2(s) + 4H^+(aq) + SO_4^{2-}(aq) + 2e^- \rightarrow PbSO_4(s) + 2H_2O(l)$

When charging: $PbSO_4(s) + 2H_2O(l) \rightarrow PbO_2(s) + 4H^+(aq) + SO_4^{2-}(aq) + 2e^-$

The nickel–cadmium cell

The nickel–cadmium (nicad) cell is the most common rechargeable cell in everyday use. Nicads produce a potential difference of 1.2 V. The anode is made from cadmium and the cathode from nickel(III) hydroxide. The electrolyte is potassium hydroxide.

Anode reaction

Under discharge: $Cd(s) + 2OH^-(aq) \rightarrow Cd(OH)_2(s) + 2e^-$

When charging: $Cd(OH)_2(s) + 2e^- \rightarrow Cd(s) + 2OH^-(aq)$

Cathode reaction

Under discharge: $Ni(OH)_3(s) + e^- \rightarrow Ni(OH)_2(s) + OH^-(aq)$
When charging: $Ni(OH)_2(s) + OH^-(aq) \rightarrow Ni(OH)_3(s) + e^-$

Lithium-ion batteries

Lithium-ion batteries are used in portable devices such as mobile phones and laptops. They have a more complicated charging system than the other cells described on this page and have a much higher potential difference, typically in the region of 3.7 V. The anode is made of graphite and the cathode is made from lithium cobalt oxide, $LiCoO_2$.

Anode reaction

Under discharge: $LiC_6(s) \rightarrow Li^+(aq) + 6C(s) + e^-$
When charging: $Li^+(aq) + 6C(s) + e^- \rightarrow LiC_6(s)$

Cathode reaction

Under discharge: $LiCoO_2(s) + Li^+(aq) + e^- \rightarrow Li_2CoO_2(s)$
When charging: $Li_2CoO_2(s) \rightarrow LiCoO_2(s) + Li^+(aq) + e^-$

The hydrogen–oxygen fuel cell

Fuel cells convert the chemical energy in a fuel such as hydrogen or methanol directly into electrical energy. The fuel undergoes an oxidation reaction with oxygen from the air using electrochemical reactions in the fuel cell.

The hydrogen–oxygen fuel cell contains two flat electrodes, each coated on one side by a thin layer of platinum catalyst. A proton exchange membrane is placed between the two electrodes. Hydrogen gas flows to the anode and air to the cathode. Water vapour, which is the reaction product, is pushed out by the stream of air.

Anode reaction

The oxidation of hydrogen is catalysed by the platinum:
$H_2(g) \rightarrow 2H^+(aq) + 2e^-$
The electrons are released into the external circuit while the H^+ ions flow through the membrane to the cathode.

Cathode reaction

Oxygen reacts with the H^+ ions and is reduced to water vapour. $4H^+(aq) + 4e^- + O_2(g) \rightarrow 2H_2O(g)$

Benefits and risks of electrochemical cells

Commercial cells are very convenient portable sources of electricity, reducing the need for expensive cabling and providing electricity to remote places.

type of cell	benefits	risks
Non-rechargeable cells	Cheap Manufactured in many sizes	Gradually run down with use Are thrown away
Rechargeable cells	Reduce the total number of batteries thrown away Vital to running of cars	Cadmium is toxic so cannot be disposed of in landfill Pb and $PbSO_4$ are toxic
Fuel cells	Only waste product is water Energy produced without fossil fuels	Hydrogen is highly flammable, so difficult to store and handle safely

Cross-section through a typical fuel cell

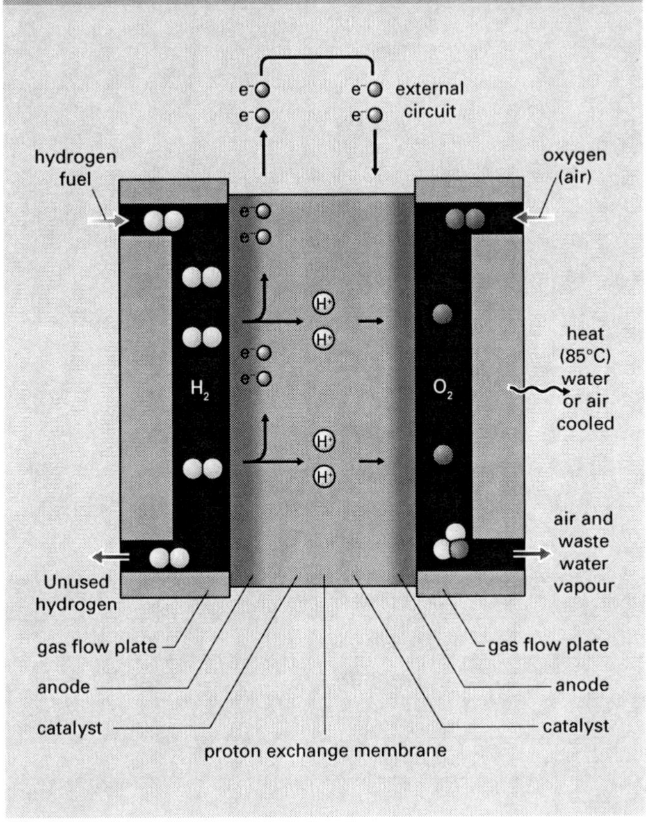

Construction of a lead–acid battery

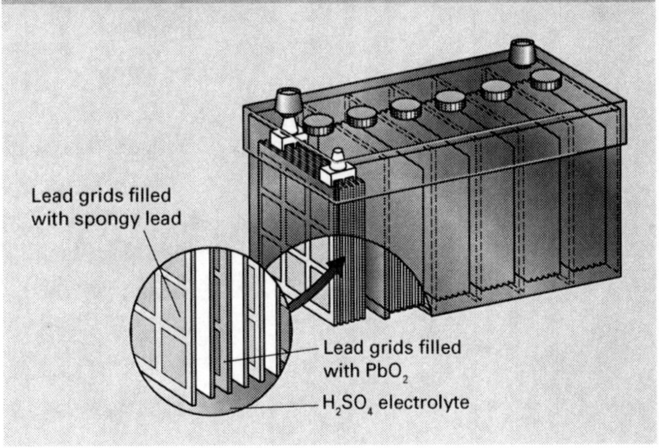

Questions

1 Write out each the anode–cathode reactions for each type of cell five times, then test to see if you can recall them from memory. Keep working through this until you can.

2 Write a two-sentence summary explaining the advantages of using a fuel cell to power a car.

5.10 The transition metals

OBJECTIVES

By the end of this section you should be able to:

- *define a transition metal*
- *write electron configurations for transition metals and their ions*
- *state the characteristics of transition metals*

Transition metals

Transition metals are elements that can form at least one stable ion with an incomplete d sub-level.

The **transition metals** are in the d block of the periodic table. It is important that you understand the distinction between a **d block element** and a transition metal. Transition metals are elements with an incomplete d sub-level that can form at least one stable ion with an incomplete d sub-level. Using this definition the first row transition elements are titanium to copper.

- Scandium has an incomplete d sub-level but forms Sc^{3+} ions which have no d sub-level electrons.
- Zinc has a complete d sub-level and forms Zn^{2+} ions with a complete d sub-level.
- As a result neither scandium nor zinc is a transition element.

Typical properties of the transition elements include:

- catalytic activity
- variable oxidation states
- the ability to form complexes
- the ability to form coloured ions

The transition metals are also stronger, more dense, and less reactive than group 1 and 2 metals.

Electron configurations

The electron configurations of the transition metals are often written using an argon core, [Ar]. This represents the electron configuration $1s^2\ 2s^2\ 2p^6\ 3s^2\ 3p^6$. Remember that the 4s shell is filled before the 3d shell and note that copper and chromium are exceptions.

element	symbol	electron configuration
titanium	Ti	[Ar] $3d^2\ 4s^2$
vanadium	V	[Ar] $3d^3\ 4s^2$
chromium	Cr	[Ar] $3d^5\ 4s^1$
manganese	Mn	[Ar] $3d^5\ 4s^2$
iron	Fe	[Ar] $3d^6\ 4s^2$
cobalt	Co	[Ar] $3d^7\ 4s^2$
nickel	Ni	[Ar] $3d^8\ 4s^2$
copper	Cu	[Ar] $3d^{10}\ 4s^1$

Chromium and copper

In chromium the electron configuration is [Ar] $3d^5\ 4s^1$ rather than [Ar] $3d^4$ $4s^2$. One of the 4s electrons is promoted to the 3d sub-level as this half-fills both the 4s and 3d sub-levels.

In copper the electron configuration is [Ar] $3d^{10}\ 4s^1$ rather than [Ar] $3d^9\ 4s^2$. One of the 4s electrons is promoted to the 3d sub-level. Again the 4s sub-level is half-filled.

Formation of ions

When the transition metals in period 4 form ions, they lose electrons from their 4s sub-level first then from their 3d sub-level. Work through the electron configurations below for some transition metal ions and make sure you can write them.

ion	electron configuration
Ti^{2+}	[Ar] $3d^2$
Cu^+	[Ar] $3d^{10}$
Fe^{3+}	[Ar] $3d^5$
Mn^{2+}	[Ar] $3d^5$

The ions of transition metals usually have a colour and often exist as **complexes**.

Variable oxidation states

The transition metals have variable oxidation states. This is possible because the d and s sub-levels are at similar energy levels. The table below shows the possible oxidation states of each element. The most important ones are shown in bold. Note that the maximum possible oxidation states increase to manganese then decrease again.

			+ 7				
		+ 6	+ 6	+ 6			
	+ 5						
+ 4	**+ 4**						
+ 3	+ 3	**+ 3**	+ 3	**+ 3**	**+ 3**	+ 3	
+ 2	+ 2	+ 2	**+ 2**	**+ 2**	**+ 2**	**+ 2**	**+ 2**
							+ 1
Ti	**V**	**Cr**	**Mn**	**Fe**	**Co**	**Ni**	**Cu**

Uses of transition elements

Transition metals and their compounds are often used as catalysts. For example, vanadium(V) oxide is used in the contact process for manufacturing sulfuric acid.

Iron is used in the Haber process to manufacture ammonia.

The transition metals titanium to copper

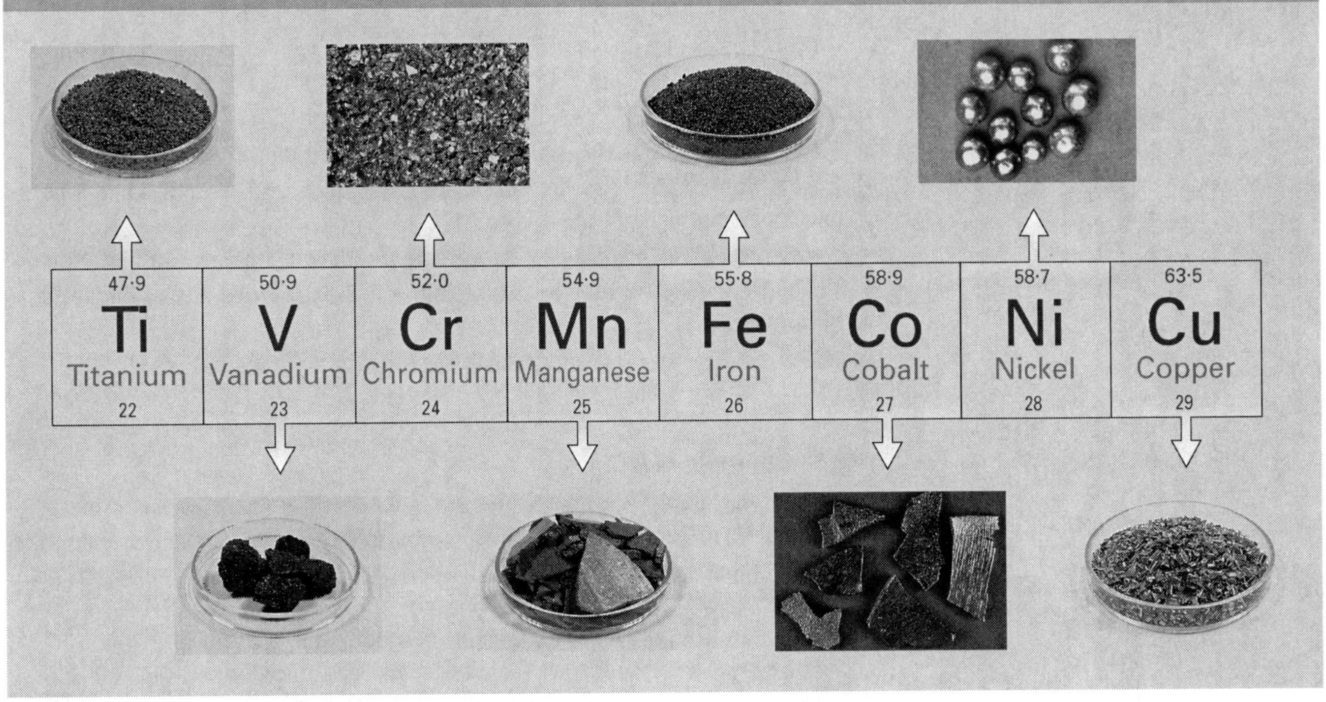

Questions

1 Write electron configurations for the following atoms and ions:

a Cr

b Ni

c Cu^{2+}

d Ni^{2+}

e V^{3+}

2 Write out the common properties of the transition metals four times then check that you can recall them.

5.11 Complex ions and their shapes

OBJECTIVES

By the end of this section you should:

- *be able to define the term ligand*
- *know that co-ordinate bonding is involved in complex formation*
- *know the co-ordination number and shapes of common complex ions*
- *know how to name common complex ions*
- *know that cisplatin is used as an anticancer drug*

A **complex** is a central metal ion surrounded by **ligands**. The ligand donates a pair of electrons to the central metal ion. You must know the formulae of common ligands and the structures of the complex ions that they form.

Ligands

Ligands can be negative ions or uncharged molecules that have one or more lone pairs of electrons to donate.

Examples of ligands include:

- halide ions such as chloride ions, Cl^-
- cyanide ions, CN^-
- hydroxide ions, OH^-
- ammonia, NH_3
- water, H_2O

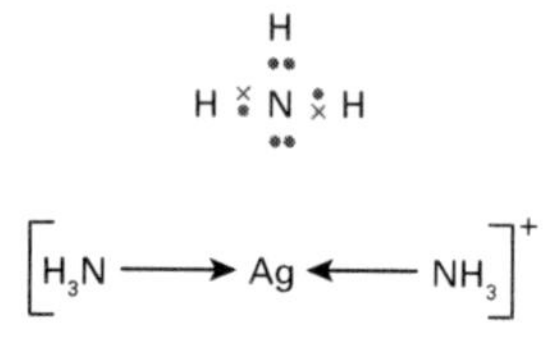

Unidentate ligands

A **unidentate** ligand has one pair of electrons that it can donate.

- common unidentate ligands include: Cl^-, CN^-, NH_3, H_2O
- the donation of the pair of electrons to the complex ion is shown as an arrow pointing from the ligand to the metal ion.
- in the diamminesilver(I) complex ion each ammonia molecule donates a lone pair of electrons to the silver ion. This ion is commonly known as Tollens' reagent and is used to test for aldehydes.

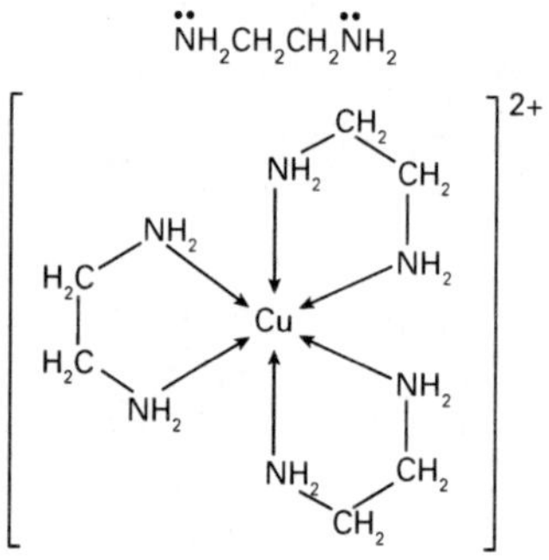

Bidentate ligands

A **bidentate** ligand has two pairs of electrons that it can donate.

- 1,2-diaminoethane is a bidentate ligand as each nitrogen atom has a lone pair of electrons.
- Each molecule can form two coordinate bonds with the central metal ion.
- The abbreviation en may be used to show 1,2-diaminoethane in diagrams of complexes.

Another example of a bidentate ligand is the ethanedioate ion, $C_2O_4^{2-}$.

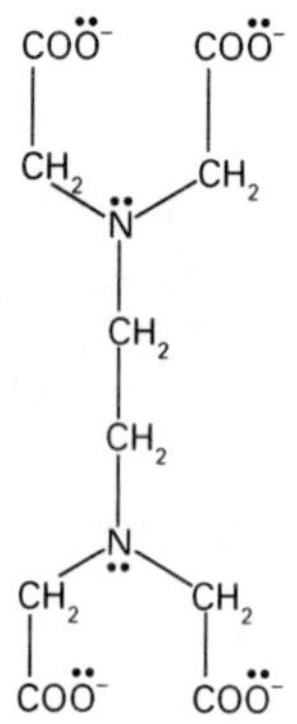

Multidentate ligands

Multidentate ligands are able to donate more than two pairs of electrons. The ligand $EDTA^{4-}$, shown on the left, can donate six pairs of electrons to the central metal ion. It has two lone pairs on nitrogen atoms, and four on oxygen atoms in the carboxylate groups.

Haemoglobin, found in red blood cells, is a protein that binds to oxygen allowing blood to carry it round the body. Each haemoglobin molecule is made from four smaller sub-units, each of which contains a haem group. This is a complex involving an iron(II) ion and a multidentate ligand that forms four co-ordinate bonds with a central metal ion.

Octahedral complexes

Complexes that contain six co-ordinate bonds have an **octahedral** shape. They have a co-ordination number of 6 (this is the number of pairs of electrons donated to the central metal ion). Examples of octahedral complexes include $[Cu(H_2O)_6]^{2+}$, $[V(H_2O)_6]^{3+}$ and $[Cu(H_2O)_4(OH)_2]$

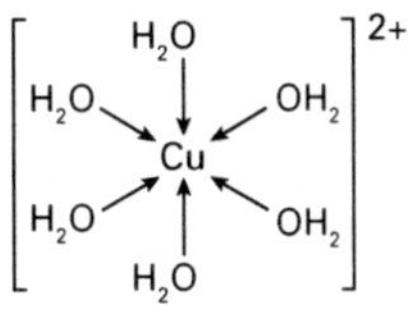

Tetrahedral complexes

Complexes containing chloride or cyanide ions have only four co-ordinate bonds since the chloride and cyanide ligands are large. They have a tetrahedral shape and a co-ordination number of 4. Examples of tetrahedral complexes include $[CuCl_4]^{2-}$ and $[FeCl_4]^{2-}$.

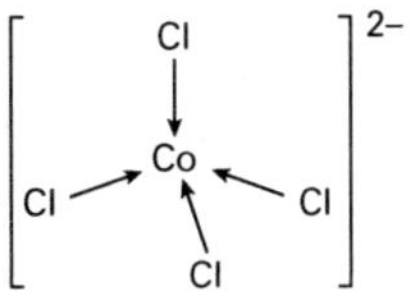

Square planar complexes

Some complexes with four ligands are square planar instead of tetrahedral. The tetracyanonickelate(II) ion. $[Ni(CN)_4]^{2-}$ is square planar.

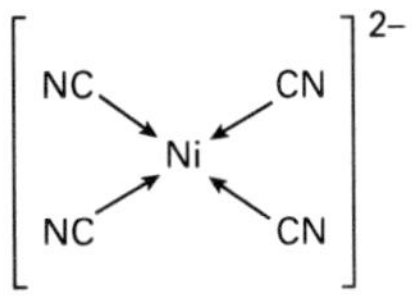

Cisplatin

The complex diamminedichloroplatinum(II), $[PtCl_2(NH_3)_2]$, is square planar and exists as *cis* and *trans* steroisomers.

- The *cis* isomer has two ammine ligands next to each other.
- the *trans* isomer has two ammine ligands on opposite sides.

The *cis* isomer is commonly called cisplatin and is used as an anticancer drug.

- It is used to treat cancer of the ovaries, testes, bladder, stomach, and lungs.
- The side effects of cisplatin include nausea, vomiting, allergic reactions, hearing loss, and kidney problems.
- It is not entirely clear how the drug works but it is thought to be converted in cells into a reactive ion that binds to DNA and stops cancer cells dividing.

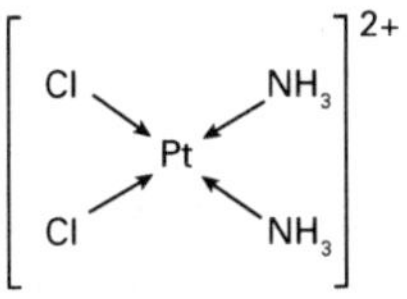

*cis*platin

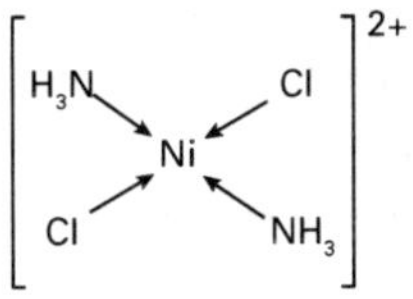

*trans*platin

Questions

1 State the co-ordination number and oxidation states of the central metal ion in:

a $[Cu(H_2O)_2(NH_3)_4]^{2+}$

b $[FeCl_4]^{2-}$

c $[Ag(CN)_2]^{-}$

2 Name and predict the shapes of each of the complexes in question 1.

Ligands and co-ordinate bonds

A ligand is an ion or molecule that can donate a pair of electrons to a central metal ion, forming a covalent bond.

A co-ordinate bond is a covalent bond in which one atom donates both electrons.

Naming complexes

Work through the rules below making sure you can see how the following complex ions are named.

hexaaquacopper(II) $[Cu(H_2O)_6]^{2+}$

tetraachlorocobaltate(II) $[CoCl_4]^{2-}$

diamminesilver(I) $[Ag(NH_3)_2]^{+}$

tetraamminediaquacopper(II) $[Cu(H_2O)_2(NH_3)_4]^{2+}$

ligand	name in complex
water	aqua
ammonia	ammine
chloride	chloro
hydroxide	hydroxo
cyanide	cyano

metal	name in anionic complexes
chromium	chromate
cobalt	cobaltate
copper	cuprate
iron	ferrate
silver	argentate

A roman numeral in brackets shows the oxidation state of the metal.

Anionic complexes have a name ending in –ate, cationic complexes end with the normal name of the metal.

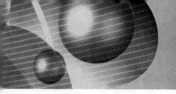

5.12 Formation of coloured ions

OBJECTIVES

By the end of this section you should know:

- *that transition ions can be identified by their colour*
- *that colour arises from electronic transitions*
- *that colour changes arise from changes in oxidation state and/or co-ordination number*
- *that colorimetry can be used to determination the concentration of coloured ions*
- *the colours of the different complex ions of chromium and cobalt*

A **complex** which is observed as having a colour is absorbing some of the colours in white light. The colours that are observed are complementary colours, which are those that remain after absorption of other colours by a substance.

Electronic transitions

Electrons are able to undergo electronic transitions. These are changes from one energy level to another.

- Electrons occupy particular energy levels in atoms.
- When an atom absorbs energy, an electron can be promoted from its normal ground state to a higher energy level.
- The electron is then in a higher energy state called an excited state.
- Energy is emitted when the electron returns to the ground state.
- The frequency of the energy absorbed or emitted is given by Planck's equation.

$$\Delta E = h\nu$$

ΔE = difference in energy in joule, J

h = Planck's constant ($6{\cdot}626 \times 10^{-34}$ J s)

ν = frequency of the electromagnetic radiation in hertz, Hz

You are not expected to complete any calculations using this equation but should know that

- a large energy transition involves high frequency radiation (in the blue end of the spectrum)

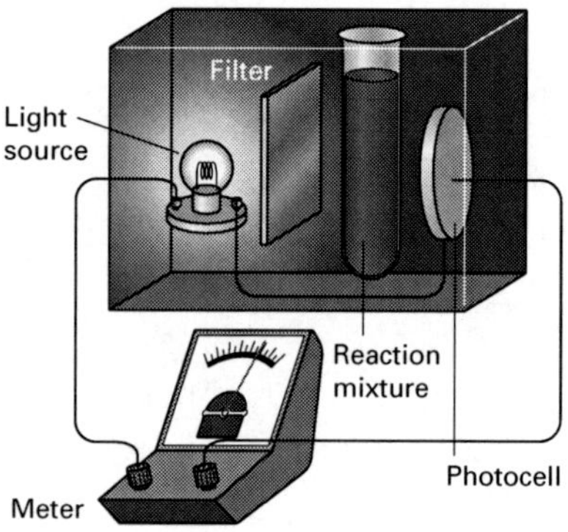

Colorimetry

Light is absorbed as it passes through a solution of coloured ions. The amount of light absorbed by an aqueous solution of ions is dependent on the concentration of the solution. A colorimeter can be used to measure this absorbance and hence determine the concentration of a solution.

Here is a simple explanation of how the colorimeter works.

- The sample is placed in a cuvette.
- White light is passed through a filter and into the sample.
- The filter is the complementary colour to that of the solution (this enables maximum absorbance by the coloured solution).
- The photocell detects the amount of light transmitted through the sample and hence the absorbance of light by the sample.

A standard curve is obtained using solutions of known concentrations. The absorbance of the unknown solution is then compared against this curve.

d-to-d transitions

In an unbonded transition metal or ion the five d orbitals are at the same energy level. When ligands are bonded to the transition metal, the d sub-level splits into two slightly different energy levels. This allows electron transitions to occur between the energy levels. A complex is coloured if this absorption corresponds to a frequency of visible light. Since space is needed in a higher-energy d orbital to accept the promoted electron an ion must have a partially filled d sub-level to be coloured.

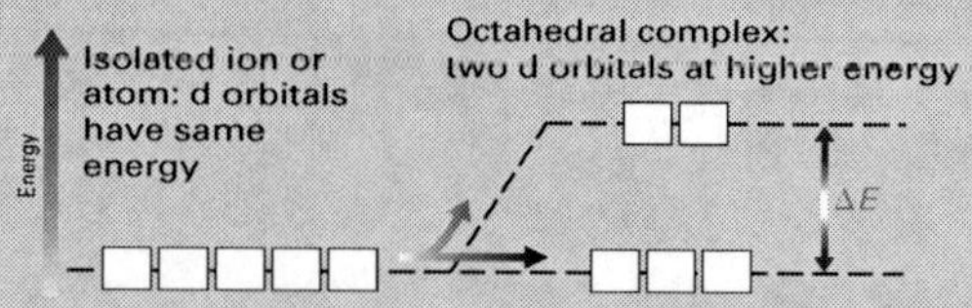

Three factors influence the colours of transition metals:

1: Oxidation state

The higher the oxidation state of the metals ions the greater the amount of d sub-level splitting. This means more energy needs to be absorbed to promote an electron and therefore alters colour.

(continued next column)

Ligand

Different ligands cause different amounts of d sub-level splitting.

The splitting in general increases in the order $Cl^- < OH^- < H_2O < NH_3 < CN^-$

Co-ordination number

The amount of d sub-level splitting is greater in an octahedral than in a tetrahedral complex.

Complex ions of chromium

Aqueous potassium dichromate(VI), $K_2Cr_2O_7$, acidified with dilute sulfuric acid is used as an oxidizing agent. It can be used to oxidize primary and secondary alcohols and in redox titrations. It is useful in both of these reactions as the colour of the chromium changes as its oxidation number changes.

Make sure you can remember the equations and colours of the ions for the reduction of dichromate(VI) and for the oxidation of chromium(III). Read the text below alongside the boxes containing the formulae of the ions.

Cr(VI) is reduced to Cr(II) using a mixture of zinc and hydrochloric acid.

- Initially $Cr_2O_7^{2-}$ is reduced to $[CrCl_2(H_2O)_4]^+$ ions.
- Then the $[CrCl_2(H_2O)_4]^+$ ions are reduced to $[Cr(H_2O)_6]^{2+}$ ions.
- This stage is carried out in the absence of air as the hexaaquachromium(II) ion is a powerful reducing agent and is rapidly oxidized in the presence of air to green $[CrCl_2(H_2O)_4]^+$ ions.

$[Cr(H_2O)_6]^{3+}(aq)$ is oxidized using hydrogen peroxide in alkaline solution. This process occurs in three steps:

- Addition of NaOH(aq) $[Cr(H_2O)_6]^{3+}(aq) + 3OH^-(aq) \rightarrow [Cr(H_2O)_3(OH)_3](s) + 3H_2O(l)$
- Addition of excess NaOH(aq) $[Cr(H_2O)_3(OH)_3](s) + 3OH^-(aq) \rightarrow [Cr(OH)_6]^{3-}(aq) + 3H_2O(l)$
- Warm with $H_2O_2(aq)$ $2[Cr(OH)_6]^{3-}(aq) + 3H_2O_2(aq) \rightarrow 2CrO_4^{2-}(aq) + 2OH^-(aq) + 8H_2O(l)$

Complex ions of chromium

oxidation state	complex	colour
+6	$Cr_2O_7^{2-}$	orange
+3	$[CrCl_2(H_2O)_4]^+$	green
+2	$[Cr(H_2O)_6]^{2+}$	blue

oxidation state	complex	colour
+3	$[Cr(H_2O)_6]^{3+}(aq)$	ruby
+3	$[Cr(H_2O)_3(OH)_3](s)$	grey-green ppt.
+3	$[Cr(OH)_6]^{3-}(aq)$	dark green
+6	$CrO_4^{2-}(aq)$	yellow

Complex ions of cobalt

Cobalt(II) complexes can be oxidized to cobalt(III) complexes in a similar way to the complexes of chromium. Work through the equations below using the boxes showing the formulae as before.

Cobalt(II) can be oxidized to cobalt(III) in alkaline solution. This reaction takes place in two steps using hydrogen peroxide in alkaline solution:

- Addition of NaOH(aq) $[Co(H_2O)_6]^{2+}(aq) + 2OH^-(aq) \rightarrow [Co(H_2O)_4(OH)_2](s) + 2H_2O(l)$
- Warm with $H_2O_2(aq)$ $2[Co(H_2O)_4(OH)_2](s) + H_2O_2(aq) \rightarrow 2[Co(H_2O)_3(OH)_3](s) + 2H_2O(l)$

Cobalt(II) in ammoniacal solution can be oxidized by air. This process occurs in three steps

- Addition of $NH_3(aq)$ $[Co(H_2O)_6]^{2+}(aq) + 2NH_3(aq) \rightarrow [Co(H_2O)_4(OH)_2](s) + 2NH_4^+(aq)$
- Addition of excess $NH_3(aq)$ $[Co(H_2O)_4(OH)_2](s) + 6NH_3(aq) \rightarrow [Co(NH_3)_6]^{2+}(aq) + 4H_2O(l) + 2OH^-(aq)$
- Oxidation in air $[Co(NH_3)_6]^{2+}(aq) \rightarrow [Co(NH_3)_6]^{3+}(aq) + e^-$
- This complex ion is actually yellow but other complexes are formed at the same time, turning the mixture brown.

Cobalt complexes and their colours

oxidation state	complex	colour
+2	$[Co(H_2O)_6]^{2+}(aq)$	pink
+2	$[Co(H_2O)_4(OH)_2](s)$	blue-green ppt.
+3	$[Co(H_2O)_3(OH)_3](s)$	dark brown ppt.

oxidation state	complex	colour
+2	$[Co(NH_3)_6]^{2+}(aq)$	pale brown
+3	$[Co(NH_3)_6]^{3+}(aq)$	yellow

Questions

1. Write a two-sentence explanation of why hexaaquacopper(II) ions appear coloured in solution.
2. Make a flow diagram showing the colours of the ions of chromium and cobalt mentioned on the spread. Then cover up sections of the diagram and practise writing them out until you can remember them.

5.13 Redox titrations

OBJECTIVES

By the end of this section you should:

- *have reviewed the mole relationships used in other parts of the course*
- *understand the redox titration of Fe^{2+} with MnO_4^- and $Cr_2O_7^{2-}$ in acid solution*
- *be able to recall the colours of some important ions used in redox titrations*

In AS chemistry and Unit 4 you will have learnt about acid–base titrations. **Redox titrations** are carried out using the same equipment and are similar in principle to acid–base titrations but rely on redox reactions. The most common oxidizing agents used in redox titrations are the manganate(VII) ion and the dichromate(VI) ion. The reducing agent thiosulfate, $S_2O_3^{2-}$, can also be used.

Manganate(VII) in titrations

Acidified aqueous potassium manganate(VII), $KMnO_4$, can act as an oxidizing agent in redox titrations. Make sure you can write and balance the half-equation for the reduction of the manganate(VII) to the manganese(II) ion:

$$MnO_4^-(aq) + 8H^+(aq) + 5e^- \rightarrow Mn^{2+}(aq) + 4H_2O(l)$$

The hydrogen ions are supplied by dilute sulfuric acid. The most common use of manganate(VII) in redox titrations is to find the concentration of aqueous iron(II) ions. The iron(II) ions are oxidized to iron(III).

$$Fe^{2+}(aq) \rightarrow Fe^{3+}(aq) + e^-$$

When the two half-equations are combined the overall equation is:

$$MnO_4^-(aq) + 8H^+(aq) + 5Fe^{2+}(aq) \rightarrow Mn^{2+}(aq) + 4H_2O(l) + 5Fe^{3+}(aq)$$

When these titrations are carried out:

- Aqueous iron(II) ions are placed in the conical flask.
- Dilute sulfuric acid is added.
- Aqueous potassium manganate(VII) is added from a burette to the flask.
- The manganate(VII) loses its colour as it enters the iron(II) solution.
- The endpoint is when a permanent pale pink tinge caused by excess manganate(VII) is first seen.

Worked example

Calculate the concentration of a solution of potassium manganate(VII) if 20.0 cm^3 of a solution of 0.10 mol dm^{-3} iron(II) sulfate reacts exactly with 23.4 cm^3 of potassium manganate(VII).

No of moles of iron(II) sulfate = $0.1 \times (20.0/1000) = 2 \times 10^{-3}$

No of moles of potassium manganate = $2 \times 10^{-3} \div 5 = 4 \times 10^{-4}$

Concentration of potassium manganate = $4 \times 10^{-4} \div (23.4/1000) = 0.017$ mol dm^{-3}

Dichromate(VI) in titrations

Acidified aqueous potassium dichromate(VI), $K_2Cr_2O_7$, can also be used as an oxidizing agent in redox titrations. The titration is carried out in the same way as with aqueous potassium manganate(VII).

The relevant half-equations and overall equations are:

$$Cr_2O_7^{2-}(aq) + 14H^+(aq) + 6e^- \rightarrow 2Cr^{3+}(aq) + 7H_2O(l)$$

$$Fe^{2+}(aq) \rightarrow Fe^{3+}(aq) + e^-$$

$$Cr_2O_7^{2-}(aq) + 14H^+(aq) + 6Fe^{2+}(aq) \rightarrow 2Cr^{3+}(aq) + 7H_2O(l) + 6Fe^{3+}(aq)$$

As with potassium managate(VII) this reaction is self-indicating but the colour change may be difficult to see so an indicator may be added.

Thiosulfate in titrations

If a reducing agent is needed for a redox titration then the thiosulfate ion, $S_2O_3^{2-}$, is used. On oxidation of the thiosulfate ion the tetrathionate ion, $S_4O_6^{2-}$, is formed: $2S_2O_3^{2-}(aq) \rightarrow S_4O_6^{2-}(aq) + 2e^-$

Thiosulfate is frequently used to find the concentration of aqueous iodine, which is reduced to iodide ions: $I_2(aq) + 2e^- \rightarrow 2I^-(aq)$

Overall: $2S_2O_3^{2-}(aq) + I_2(aq) \rightarrow S_4O_6^{2-}(aq) + 2I^-(aq)$

Starch suspension is used as an indicator near the endpoint. It forms a deep purple colour with iodine; the endpoint is when this colour first disappears.

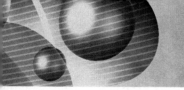

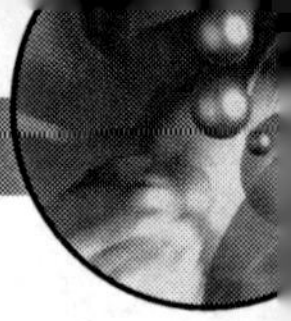

Mole relationships

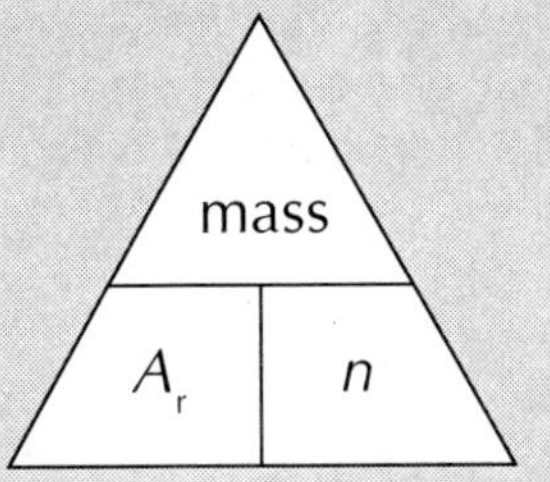

A_r = relative atomic mass

n = number of moles

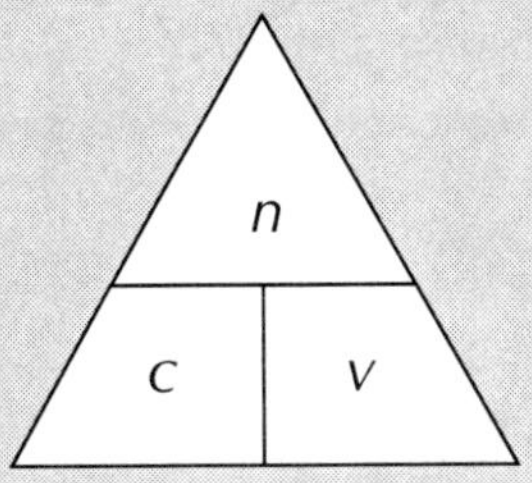

c = concentration in mol dm^{-3}

V = volume in dm^3

n = number of moles

HSW: Which acid?

Sulfuric acid is suitable for both dichromate(VI) and manganate(VII) titrations. However, hydrochloric acid is only suitable for dichromate(VI) titrations. This is because of the strength of manganate(VII) ions as an oxidizing agent. It is able to oxidize chloride ions to chlorine.

Colours of important ions

ion	colour in aqueous solution
MnO_4^-(aq)	purple
Mn^{2+}(aq)	very pale pink
$Cr_2O_7^{2-}$(aq)	orange
Cr^{3+}	green

Questions

1. Calculate the concentration in mol dm^{-3} of a solution of potassium dichromate(VI) if 26.4 cm^3 of 0.05 mol dm^{-3} iron(II) sulfate reacts exactly with 25.0 cm^3 of the potassium dichromate(VI).
2. What volume of 0.1 mol dm^{-3} sulfuric acid would be needed to provide enough H^+ ions for the reaction of 52.4 cm^3 of 0.2 mol dm^{-3} potassium manganate(VII) with iron(II) ions.
3. Write out the half-equations for the following reactions:
   - **a** manganate(VII) as an oxidizing agent
   - **b** dichromate(VI) as an oxidizing agent
   - **c** thiosulfate as a reducing agent

5.14 Uses of transition metals and their ions

OBJECTIVES

By the end of this section you should:

- *know that transition metals can act as heterogeneous and homogeneous catalysts*
- *understand the use of a support medium to maximize surface area and minimize cost*
- *understand how V_2O_5 acts as a catalyst in the contact process*
- *know that Fe can be used as a catalyst in the Haber process*
- *know that catalysts can be poisoned by impurities*
- *know that a Cr_2O_3 catalyst is used in the manufacture of methanol*

A **catalyst** is a substance that increases the rate of a chemical reaction by providing an alternative reaction route with a lower activation energy. Transition metals and their compounds can act as catalysts.

Common features of catalysts

Catalysts may catalyse only one reaction or a whole class of similar reactions. All catalysts have some features in common.

- Catalysts are often only needed in small amounts.
- Solid catalysts are more effective when used as a thin layer or as a powder as this increases their surface area.
- For an equilibrium reaction, catalysts increase the rate of the forward reaction and the reverse reaction by the same ratio.
  As a result catalysis do not alter the position of equilibrium or the value of the equilibrium constant, K_c.

Heterogeneous catalysts

Heterogeneous catalysts exist in a different state or phase to the reactants in the reaction. You will be familiar with examples of heterogeneous catalysts from AS level. Reactions occur at the surface of the catalyst in regions called active sites.

- Reactants are **adsorbed** to the active sites.
- The reaction takes place.
- Products are **desorbed** from the active site.

Catalysts are important in industrial processes. For example, methanol is manufactured from carbon monoxide and hydrogen using a catalyst containing chromium(III) oxide, Cr_2O_3.

$$CO(g) + 2H_2(g) \rightarrow CH_3OH(l)$$

The carbon monoxide and hydrogen are manufactured by the reaction of coke with steam.

The Haber process

The Haber process is an important industrial process as artificial fertilizers are made from the ammonia produced. The ammonia is manufactured from nitrogen and hydrogen in the presence of an iron catalyst.

- The overall equation is $N_2(g) + 3H_2(g) \rightleftharpoons 2NH_3(g)$
- Nitrogen molecules and hydrogen molecules are adsorbed onto the surface of the iron catalyst.
- N≡N and H–H bonds break and N–H bonds are formed.
- Ammonia molecules desorb from the surface.

The hydrogen needed is obtained by the reaction of methane with steam

$$CH_4(g) + 2H_2O(g) \rightarrow CO_2(g) + 4H_2(g)$$

Natural gas is the source of the methane needed. This often contains hydrogen sulfide, H_2S, as an impurity. This can act as a poison to the iron catalyst by adsorbing strongly to the active sites. The natural gas is treated before use to remove the sulfur.

The contact process

The use of vanadium(V) oxide in the contact process for the manufacture of sulfuric acid relies on the variable oxidation states of vanadium. Work through the equations below checking that you can work out the oxidation state of vanadium in each one.

In the second stage of sulfuric acid manufacture, sulfur(IV) oxide is oxidized to sulfur(VI) oxide. This takes place in two steps:

Overall $SO_2(g) + \frac{1}{2}O_2(g) \rightleftharpoons SO_3(g)$

Step 1 $SO_2(g) + V_2O_5(s) \rightarrow SO_3(g) + V_2O_4(s)$

Step 2 $V_2O_4(s) + \frac{1}{2}O_2(g) \rightarrow V_2O_5(s)$

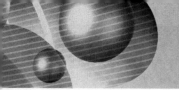

Homogeneous catalysts

Homogeneous catalysts are in the same state or phase as the reactants. Their catalytic action involves variable oxidation states. Two key examples of homogeneous catalysts are given below.

Iodide ions and peroxodisulfate ions

Aqueous peroxodisulfate ions, $S_2O_8^{2-}$, are able to oxidize aqueous iodide ions to iodine. This reaction takes place slowly because both of the reactant ions are negatively charged and therefore tend to repel each other rather than colliding. Aqueous iron(II) ions catalyze the reaction by acting as an intermediate in the transfer of electrons from iodide ions to peroxodisulfate ions:

Overall $S_2O_8^{2-}(aq) + 2I^-(aq) \rightleftharpoons 2SO_4^{2-}(aq) + I_2(aq)$

Step 1 $S_2O_8^{2-}(aq) + 2Fe^{2+}(aq) \rightarrow 2SO_4^{2-}(aq) + 2Fe^{3+}(aq)$

Step 2 $2Fe^{3+}(aq) + 2I^-(aq) \rightarrow 2Fe^{2+}(aq) + I_2(aq)$

Note that the Fe(II) ions are regenerated in Step 2 and are therefore acting as a catalyst.

Manganate(VII) ions and ethanedioate ions

When aqueous manganate(VII) is titrated with acidified aqueous ethanedioate the purple colour decolorizes slowly at first and then decolorizes immediately as more manganate(VII) is added. This happens because manganese(II) is formed in the reaction, which then catalyses the reaction. This is an example of autocatalysis.

As with the iodine and peroxodisulfate reaction, the reaction is slow because both reactant ions are negatively charged. Once the positively charged manganese(II) ions are formed, they can act as an intermediate:

Overall $2MnO_4^-(aq) + 8H^+(aq) + 5C_2O_4^{2-}(aq) \rightarrow 2Mn^{2+}(aq) + 10CO_2(g) + 4H_2O(l)$

Step 1 $MnO_4^-(aq) + 8H^+(aq) + 4Mn^{2+}(aq) \rightarrow 5Mn^{3+}(aq) + 4H_2O(l)$

Step 2 $5C_2O_4^{2-}(aq) + 5Mn^{3+}(aq) \rightarrow 10CO_2(g) + 5Mn^{2+}(aq)$

Note that the manganese(II) ions are regenerated in Step 2.

Catalyst supports

Some catalysts such as platinum and rhodium are expensive and have to be used in a very efficient way. This is often achieved by coating a thin layer of the catalyst onto an inexpensive solid support such as porous ceramic beads or honeycomb. This support has a slightly rough surface and so provides a huge surface area.

Energy level diagram

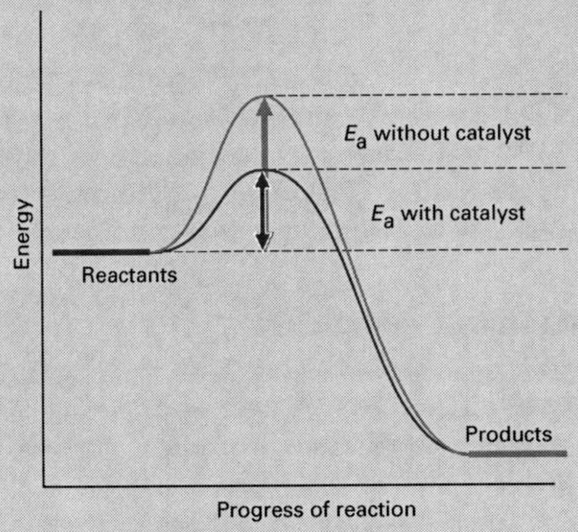

Questions

1 a Write out the equations for the second stage in the manufacture of sulfuric acid.

b Work out the oxidation numbers of the sulfur and vanadium in each of the reactions involved in the contact process.

2 Write a two-sentence explanation of heterogeneous catalysis.

3 Explain why the use of an iron catalyst in the Haber process increases the rate of attainment of equilibrium but has no impact on the yield of ammonia obtained.

5.15 Lewis acids and bases and metal–aqua ions

OBJECTIVES

By the end of this section you should:

- *be able to define a Lewis acid and a Lewis base*
- *know that metal–aqua ions are formed in aqueous solution*
- *know about some common metal–aqua ions*
- *know that metal–aqua ions can undergo hydrolysis reactions*

Lewis acid and bases

Using Brønsted–Lowry definitions:

- an acid is a proton donor
- a base is a proton acceptor

Using Lewis definitions:

- an acid is an electron pair acceptor
- a base is an electron pair donor

You will be familiar with the Brønsted–Lowry definitions for acids and bases. For transition metal chemistry it is useful to use the **Lewis** definitions. Ammonia, which is a common ligand in transition metal complexes is a base because it forms a co-ordinate bond by donating a pair of electrons to the central metal ion. As a result the metal ion is acting as a Lewis acid.

Metal–aqua ions

Transition metal ions such as Cu^{2+} and Co^{3+} have a high charge density. They strongly attract water molecules when in aqueous solution forming hexaaqua complexes of the form $[M(H_2O)_6]^{2+}$ and $[M(H_2O)_6]^{3+}$. These octahedral complexes have a co-ordination number of 6.

$[M(H_2O)_6]^{2+}$ ions

Iron(II), copper, and cobalt ions all dissolve in water to form hexaaqua complexes:

$CuSO_4(s) + 6H_2O(l) \rightarrow [Cu(H_2O)_6]^{2+}(aq) + SO_4^{2-}(aq)$

$CoCl_2(s) + 6H_2O(l) \rightarrow [Co(H_2O)_6]^{2+}(aq) + 2Cl^-(aq)$

$FeSO_4(s) + 6H_2O(l) \rightarrow [Fe(H_2O)_6]^{2+}(aq) + SO_4^{2-}(aq)$

solid	colour	complex ion	colour
$CuSO_4(s)$	white	$[Cu(H_2O)_6]^{2+}(aq)$	blue
$CoCl_2(s)$	blue	$[Co(H_2O)_6]^{2+}(aq)$	pink
$FeSO_4(s)$	pale green	$[Fe(H_2O)_6]^{2+}(aq)$	pale green

Make sure you can recall the colours of each of the ions and solids.

$[M(H_2O)_6]^{3+}$ ions

Chromium(III) and iron(III) compounds dissolve in water forming hexaaqua complexes. Aluminium (which is not a transition metal) also forms hexaaqua complexes owing to the high charge density of the Al^{3+} ion. Note that the hexaaquaaluminium complex is colourless as aluminium is not a transition metal and its electrons do not undergo d-to-d transitions.

$AlCl_3(s) + 6H_2O(l) \rightarrow [Al(H_2O)_6]^{3+}(aq) + 3Cl^-(aq)$

$CrCl_3(s) + 6H_2O(l) \rightarrow [Cr(H_2O)_6]^{3+}(aq) + 3Cl^-(aq)$

$FeCl_3(s) + 6H_2O(l) \rightarrow [Fe(H_2O)_6]^{3+}(aq) + 3Cl^-(aq)$

Make sure you can recall the colours of each ion. Note that the pale violet colour of $[Fe(H_2O)_6]^{3+}(aq)$ is not seen as the complex hydrolyses rapidly to form $[Fe(H_2O)_5(OH)]^{2+}(aq)$. The $[Cr(H_2O)_6]^{3+}(aq)$ colour is sometimes not seen, because chloride ions and other ligands may replace the water to give green complexes.

Hydrolysis reactions of complexes

It is possible for metal–aqua ions to undergo two main types of reactions. The breaking of the co-ordinate bond between an aqua ligand and the central metal ion results in **ligand substitution**. Breaking of the O–H bond in an aqua ligand releases a hydrogen ion in a **hydrolysis reaction**.

Hexaaquaaluminium complexes have a pH of approximately 3 in aqueous solution because the $[Al(H_2O)_6]^{3+}$ (aq) complex reacts with water in a hydrolysis reaction producing H_3O^+ ions. (These are formed because a water molecule acts as a base in the reaction and accepts H^+.)

$$[Al(H_2O)_6]^{3+}(aq) + H_2O(l) \rightleftharpoons [Al(H_2O)_5(OH)]^{2+}(aq) + H_3O^+(aq)$$

The position of this equilibrium lies a long way to the left hand side so the solution is only weakly acidic.

Transition metal complexes that contain a central metal ion with a 2+ or 3+ charge also undergo hydrolysis reactions.

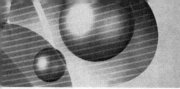

Metal ions with high charge/size ratios are very **polarizing**. They polarize the aqua ligands and weaken the O–H bonds. A water molecule from the solvent then removes a hydrogen ion from an aqua ligand forming a hydroxo ligand in the complex.

M^{3+} transition metal ions

Two examples of the hydrolysis reactions undergone by M^{3+} transition metal ions are given here. In both cases a pentaaquahydroxo complex is formed:

$$[Cr(H_2O)_6]^{3+}(aq) + H_2O(l) \rightleftharpoons [Cr(H_2O)_5(OH)]^{2+}(aq) + H_3O^+(aq)$$
$$[Fe(H_2O)_6]^{3+}(aq) + H_2O(l) \rightleftharpoons [Fe(H_2O)_5(OH)]^{2+}(aq) + H_3O^+(aq)$$

M^{2+} transition metal ions

M^{2+} transition metal ions are also able to undergo hydrolysis reactions, the positions of the equilibria established lie a long way to the left so the solutions are only weakly acidic. Some typical reactions are:

$$[Cu(H_2O)_6]^{2+}(aq) + H_2O(l) \rightleftharpoons [Cu(H_2O)_5(OH)]^{+}(aq) + H_3O^+(aq)$$
$$[Co(H_2O)_6]^{2+}(aq) + H_2O(l) \rightleftharpoons [Co(H_2O)_5(OH)]^{+}(aq) + H_3O^+(aq)$$
$$[Fe(H_2O)_6]^{2+}(aq) + H_2O(l) \rightleftharpoons [Fe(H_2O)_5(OH)]^{+}(aq) + H_3O^+(aq)$$

Colours of $[M(H_2O)_6]^{3+}$ ions

complex ion	colour
$[Al(H_2O)_6]^{3+}(aq)$	colourless
$[Cr(H_2O)_6]^{3+}(aq)$	ruby
$[Fe(H_2O)_6]^{3+}(aq)$	pale violet

Colour of $[Fe(H_2O)_6]^{3+}$

The hexaaquairon(III) complex is a pale violet but hydrolyses in water to form a small amount of the brown $[Fe(H_2O)_5(OH)]^{2+}$ which hides the pale violet colour of $[Fe(H_2O)_6]^{3+}$.

Factors affecting acidity

Acid strength increases as

- the ionic radius decreases
- the size of the charge on the ion increases

As a result the most polarizing ions are ions such as Al^{3+}, Cr^{3+}, and Fe^{3+}. These attract electron density from the oxygen atoms in the aqua ligands weakening the O–H bond so less energy is needed to release a hydrogen ion.

Questions

1 Write balanced equations showing the dissolving of copper sulfate and cobalt chloride in water.

2 Explain why the Fe^{3+} ion is more polarizing than the Fe^{2+} ion.

3 Using a suitable example, explain why metal aqua ions produce weakly acidic solutions.

5.16 Reactions of the transition metal ions

OBJECTIVES

By the end of this section you should:

- *be able to describe the reactions of $M^{2+}(aq)$ and $M^{3+}(aq)$ ions with hydroxide ions, ammonia, and carbonate ions (M^{2+} = Cu, Co and Fe, M^{3+} = Al, Cr, and Fe)*
- *know that some metal hydroxides show amphoteric character by dissolving in both acids and bases*

M^{2+} and M^{3+} precipitates and their colours

complex ion	colour
$[Cu(H_2O)_4(OH)_2](s)$	blue
$[Cu(H_2O)_4(OH)_2](s)$	blue-green
$[Fe(H_2O)_4(OH)_2](s)$	green
$[Al(H_2O)_3(OH)_3](s)$	white
$[Cr(H_2O)_3(OH)_3](s)$	grey-green
$[Fe(H_2O)_3(OH)_3](s)$	brown

Aluminium hydroxide

Aluminium hydroxide redissolves in excess sodium hydroxide solution but does not redissolve in excess ammonia solution. The transition metal complexes do redissolve in excess ammonia (see ligand substitution reactions).

Precipitates and their colours

precipitate	colour
$CuCO_3(s)$	green-blue
$CoCO_3(s)$	pink
$FeCO_3(s)$	green

You must be able to describe and explain all the reactions shown below. There are lots of equations and colours to remember, so work through them carefully. Also refer back to the spread 'Lewis acids and bases and metal–aqua ions'.

Reactions with hydroxide ions

Remember that an aqueous solution of M^{2+} transition metal ions contains an equilibrium mixture with the equilibrium lying a long way to the left hand side.

$$[M(H_2O)_6]^{2+}(aq) + H_2O(l) \rightleftharpoons [M(H_2O)_5(OH)]^{+}(aq) + H_3O^{+}(aq)$$

Addition of hydroxide ions removes the H_3O^+ ions and moves the equilibrium to the right hand side until an insoluble metal(II) hydroxide precipitate forms:

$$OH^-(aq) + H_3O^+(aq) \rightarrow 2H_2O(l)$$

$$[M(H_2O)_5(OH)]^{+}(aq) + H_2O(l) \rightleftharpoons [M(H_2O)_4(OH)_2](s) + H_3O^{+}(aq)$$

So overall $[M(H_2O)_6]^{2+}(aq) + 2OH^-(aq) \rightleftharpoons [M(H_2O)_4(OH)_2](s) + 2H_2O(l)$

copper(II)	$[Cu(H_2O)_6]^{2+}(aq) + 2OH^-(aq) \rightleftharpoons [Cu(H_2O)_4(OH)_2](s) + 2H_2O(l)$
cobalt(II)	$[Co(H_2O)_6]^{2+}(aq) + 2OH^-(aq) \rightleftharpoons [Co(H_2O)_4(OH)_2](s) + 2H_2O(l)$
iron(II)	$[Fe(H_2O)_6]^{2+}(aq) + 2OH^-(aq) \rightleftharpoons [Fe(H_2O)_4(OH)_2](s) + 2H_2O(l)$

With M^{3+} transition metal ions the same series of reactions takes place starting with $[M(H_2O)_6]^{3+}(aq)$ and resulting in an insoluble metal(III) hydroxide precipitate:

$$[M(H_2O)_6]^{3+}(aq) + 3OH^-(aq) \rightarrow [M(H_2O)_3(OH)_3](s) + 3H_2O(l)$$

aluminium	$[Al(H_2O)_6]^{3+}(aq) + 3OH^-(aq) \rightleftharpoons [Al(H_2O)_3(OH)_3](s) + 3H_2O(l)$ Aluminium hydroxide redissolves to form the colourless tetrahydroxoaluminate complex: $[Al(H_2O)_3(OH)_3](s) + OH^-(aq) \rightleftharpoons [Al(OH)_4]^-(aq) + 3H_2O(l)$
chromium(III)	$[Cr(H_2O)_6]^{3+}(aq) + 3OH^-(aq) \rightleftharpoons [Cr(H_2O)_3(OH)_3](s) + 3H_2O(l)$ Chromium(III) hydroxide redissoves in excess aqueous sodium hydroxide to form the hexahydroxochromate(III) complex: $[Cr(H_2O)_3(OH)_3](s) + 3OH^-(aq) \rightleftharpoons [Cr(OH)_6]^{3-}(aq) + 3H_2O(l)$
iron(III)	$[Fe(H_2O)_6]^{3+}(aq) + 3OH^-(aq) \rightleftharpoons [Fe(H_2O)_3(OH)_3](s) + 3H_2O(l)$

Reactions with ammonia

This page shows the reactions of metal–aqua ions with ammonia acting as a base. The reactions with ammonia as a ligand are covered on the next spread.

An aqueous solution of ammonia acts as a base by reacting with the oxonium ions present in the aqueous solution of the transition metal ions:

$$NH_3(aq) + H_3O^+(aq) \rightarrow NH_4^+(aq) + H_2O(l)$$

This leads to the formation of precipitates in the same way as with the hydroxide ions:

$$[M(H_2O)_6]^{2+}(aq) + 2NH_3(aq) \rightarrow [M(H_2O)_4(OH)_2](s) + 2NH_4^+(aq)$$

copper(II)	$[Cu(H_2O)_6]^{2+}(aq) + 2NH_3(aq) \rightleftharpoons [Cu(H_2O)_4(OH)_2](s) + 2NH_4^+(aq)$
cobalt(II)	$[Co(H_2O)_6]^{2+}(aq) + 2NH_3(aq) \rightleftharpoons [Co(H_2O)_4(OH)_2](s) + 2NH_4^+(aq)$
iron(II)	$[Fe(H_2O)_6]^{2+}(aq) + 2NH_3(aq) \rightleftharpoons [Fe(H_2O)_4(OH)_2](s) + 2NH_4^+(aq)$ The green precipitate gradually turns orange-brown because iron(II)hydroxide is oxidized by air to iron(III) hydroxide

Aluminium, chromium(III), and iron(III) form precipitates in the same way.

aluminium	$[Al(H_2O)_6]^{3+}(aq) + 3NH_3(aq) \rightleftharpoons [Al(H_2O)_3(OH)_3](s) + 3NH_4^+(aq)$
chromium(III)	$[Cr(H_2O)_6]^{3+}(aq) + 3NH_3(aq) \rightleftharpoons [Cr(H_2O)_3(OH)_3](s) + 3NH_4^+(aq)$
iron(III)	$[Fe(H_2O)_6]^{3+}(aq) + 3NH_3(aq) \rightleftharpoons [Fe(H_2O)_3(OH)_3](s) + 3NH_4^+(aq)$

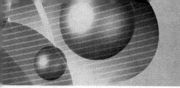

Reactions with carbonate ions

Metal–aqua ions in solution react with sodium carbonate and other carbonates. The carbonate ion acts as a Brønsted–Lowry base to produce the hydrogencarbonate ion which then reacts further to make carbon dioxide.

$$CO_3^{2-}(aq) + H_3O^+(aq) \rightleftharpoons HCO_3^-(aq) + H_2O(l)$$
$$HCO_3^-(aq) + H_3O^+(aq) \rightleftharpoons CO_2(g) + 2H_2O(l)$$
$$\text{Overall } CO_3^{2-}(aq) + 2H_3O^+(aq) \rightarrow CO_2(g) + 3H_2O(l)$$

Metal hexaaqua ions in aqueous solution are very weakly acidic. When carbonate ions are added to these solutions, metal(II) carbonates are produced (the solution is not acidic enough to form carbon dioxide). The metal(II) carbonates form precipitates.

$$[M(H_2O)_6]^{2+}(aq) + CO_3^{2-}(aq) \rightarrow MCO_3(s) + 6H_2O(l)$$

copper(II)	$[Cu(H_2O)_6]^{2+}(aq) + CO_3^{2-}(aq) \rightarrow CuCO_3(s) + 6H_2O(l)$
cobalt(II)	$[Co(H_2O)_6]^{2+}(aq) + CO_3^{2-}(aq) \rightarrow CoCO_3(s) + 6H_2O(l)$
iron(II)	$[Fe(H_2O)_6]^{2+}(aq) + CO_3^{2-}(aq) \rightarrow FeCO_3(s) + 6H_2O(l)$

Addition of carbonate ions to a solution of a M^{3+} transition metal ion does not form a metal(III) carbonate. Instead an insoluble metal(III) hydroxide is formed. $Al_2(CO_3)_3$, $Cr_2(CO_3)_3$, and $Fe_2(CO_3)_3$ do not exist.

$$2[M(H_2O)_6]^{3+}(aq) + 3CO_3^{2-}(aq) \rightleftharpoons 2[M(H_2O)_3(OH)_3](s) + 3H_2O(l) + 3CO_2(g)$$

aluminium	$2[Al(H_2O)_6]^{3+}(aq) + 3CO_3^{2-}(aq) \rightleftharpoons 2[Al(H_2O)_3(OH)_3](s) + 3H_2O(l) + 3CO_2(g)$
chromium(III)	$2[Cr(H_2O)_6]^{3+}(aq) + 3CO_3^{2-}(aq) \rightleftharpoons 2[Cr(H_2O)_3(OH)_3](s) + 3H_2O(l) + 3CO_2(g)$
iron(III)	$2[Fe(H_2O)_6]^{3+}(aq) + 3CO_3^{2-}(aq) \rightleftharpoons 2[Fe(H_2O)_3(OH)_3](s) + 3H_2O(l) + 3CO_2(g)$

Amphoteric metal hydroxides

Amphoteric substances can react with both acids and bases. Aluminium hydroxide and chromium(III) hydroxide are both amphoteric. As with all the other reactions on this page the addition of strong acid or strong base influences the position of the equilibria established when metal ions are dissolved in water.

$$[M(H_2O)_6]^{3+}(aq) + 3OH^-(aq) \rightarrow [M(H_2O)_3(OH)_3](s) + 3H_2O(l)$$

On addition of a strong acid:

$$[M(H_2O)_3(OH)_3](s) + 3H_3O^+(aq) \rightarrow [M(H_2O)_6]^{3+}(aq) + 3H_2O(l)$$

Aluminium hydroxide

Aluminium hydroxide reacts with excess aqueous sodium hydroxide to form the soluble tetrahydroxoaluminate complex and with excess acid to form the soluble hexaaquaaluminium complex:

$$[Al(H_2O)_6]^{3+}(aq) + 3OH^-(aq) \rightleftharpoons [Al(H_2O)_3(OH)_3](s) + 3H_2O(l)$$
$$[Al(H_2O)_3(OH)_3](s) + OH^-(aq) \rightleftharpoons [Al(OH)_4]^-(aq) + 3H_2O(l)$$
$$[Al(H_2O)_3(OH)_3](s) + 3H_3O^+(aq) \rightleftharpoons [Al(H_2O)_6]^{3+}(aq) + 3H_2O(l)$$

Chromium(III)

Chromium(III) ions react in a similar way to aluminium:

$$[Cr(H_2O)_6]^{3+}(aq) + 3OH^-(aq) \rightleftharpoons [Cr(H_2O)_3(OH)_3](s) + 3H_2O(l)$$
$$[Cr(H_2O)_3(OH)_3](s) + 3OH^-(aq) \rightleftharpoons [Cr(OH)_6]^{3-}(aq) + 3H_2O(l)$$
$$[Cr(H_2O)_3(OH)_3](s) + 3H_3O^+(aq) \rightleftharpoons [Cr(H_2O)_6]^{3+}(aq) + 3H_2O(l)$$

Chromate(VI) and dichromate(VI)

In aqueous solution the yellow chromate(VI) ion, $CrO_4^{2-}(aq)$, exists in equilibrium with the orange dichromate(VI) ion, $Cr_2O_7^{2-}(aq)$:

$$2CrO_4^{2-}(aq) + 2H^+(aq) \rightleftharpoons Cr_2O_7^{2-}(aq) + H_2O(l)$$

Addition of dilute sulfuric acid moves the equilibrium to the right hand side, addition of aqueous sodium hydroxide moves the equilibrium to the left hand side.

Complexes and their colours

complex ion	colour
$[Al(H_2O)_6]^{3+}(aq)$	colourless
$[Al(H_2O)_3(OH)_3](s)$	white
$[Al(OH)_4]^-(aq)$	colourless
$[Cr(H_2O)_6]^{3+}(aq)$	ruby
$[Cr(H_2O)_3(OH)_3](s)$	grey-green
$[Cr(OH)_6]^{3-}(aq)$	dark green

Questions

1. State and explain the different reactions of hexaaquacopper(II) ions with sodium hydroxide solution and carbonate ions. Use equations in your answer.
2. Explain with relevant equations the meaning of the term amphoteric.
3. Make your own summary sheet showing all the reactions and colours given on this double page.

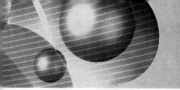

5.17 Ligand substitution reactions

OBJECTIVES

By the end of this section you should:

- *be able to explain the term ligand substitution*
- *be able to write equations for and explain ligand substitution by ammonia and chloride*
- *know that the substitution of a unidentate ligand with a bidentate or multidentate ligand leads to a more stable complex*
- *understand the effect of chelation in terms of a positive entropy change*

Ammine complexes

complex ion	colour
$[Cr(NH_3)_6]^{3+}(aq)$	purple
$[Co(NH_3)_6]^{2+}(aq)$	straw-coloured
$[Cu(NH_3)_4(H_2O)_2]^{2+}(aq)$	deep blue

Octahedral complexes

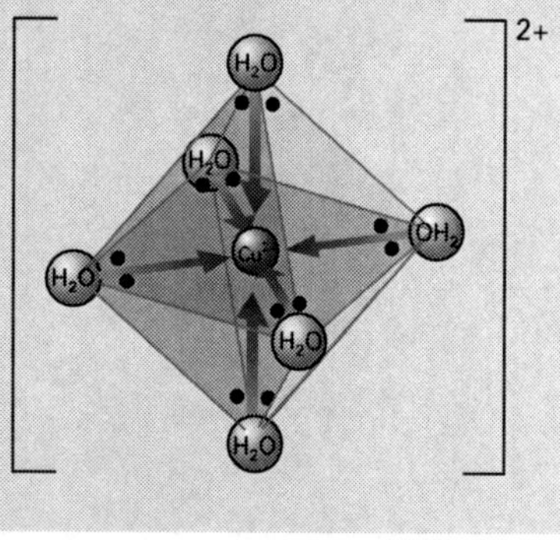

Aqua and chloro complexes

complex ion	colour
$[Cr(H_2O)_6]^{3+}(aq)$	ruby
$[Co(H_2O)_6]^{2+}(aq)$	pink
$[CoCl_4]^{2-}(aq)$	blue
$[Cu(H_2O)_6]^{2+}(aq)$	blue
$[CuCl_4]^{2-}(aq)$	olive

In the previous spread you reviewed the reactions of hexaaqua complexes when the O–H bond in the aqua ligand breaks. If the co-ordinate bond between the aqua ligand and the central metal ion breaks then a **ligand substitution** reaction takes place. You need to be able to write equations and state the colour changes for all the ligand substitution reactions given on this spread.

Substitution by ammonia

Water and ammonia are both neutral unidentate ligands. Ammonia molecules are able to replace water molecules in ligand substitution reactions. As water and ammonia have a similar size, the coordination number remains 6 and the complexes are octahedral in shape. The complex will be a different colour as the environment of the transition metal ion and therefore the splitting of the d orbitals will be different.

- All of the water ligands can be replaced but the actual number varies depending on the transition metal ion involved.
- If all six ligands are replaced then a hexaammine complex is formed and the overall equation is:

$$[M(H_2O)_6]^{3+}(aq) + 6NH_3(aq) \rightleftharpoons [M(NH_3)_6]^{3+}(aq) + 6H_2O(l)$$

Chromium(III)

With a small amount of aqueous ammonia a grey-green precipitate of chromium(III) hydroxide is formed:

$$[Cr(H_2O)_6]^{3+}(aq) + 3NH_3(aq) \rightleftharpoons [Cr(H_2O)_3(OH)_3](s) + 3NH_4^+(aq)$$

With excess aqueous ammonia, a purple complex is formed:

$$[Cr(H_2O)_6]^{3+}(aq) + 6NH_3(aq) \rightarrow [Cr(NH_3)_6]^{3+}(aq) + 6H_2O(l)$$

Cobalt(II)

Similarly two reactions are possible with aqueous cobalt(II) ions:

$$[Co(H_2O)_6]^{2+}(aq) + 2NH_3(aq) \rightleftharpoons [Co(H_2O)_4(OH)_2](s) + 2NH_4^+(aq)$$

$$[Co(H_2O)_6]^{2+}(aq) + 6NH_3(aq) \rightarrow [Co(NH_3)_6]^{2+}(aq) + 6H_2O(l)$$

Note that the complex formed readily oxidizes in air forming a dark brown mixture containing aqueous hexamminecobalt(III):

$$[Co(NH_3)_6]^{2+}(aq) \rightarrow [Co(NH_3)_6]^{3+}(aq) + e^-$$

Copper(II)

With copper(II) ions the reaction with excess ammonia results in the ligand substitution of four aqua ligands, forming a complex with four ammine ligands in a square planar arrangement with an aqua ligand above and below the plane:

$$[Cu(H_2O)_6]^{2+}(aq) + 2NH_3(aq) \rightleftharpoons [Cu(H_2O)_4(OH)_2](s) + 2NH_4^+(aq)$$

$$[Cu(H_2O)_6]^{2+}(aq) + 4NH_3(aq) \rightarrow [Cu(NH_3)_4(H_2O)_2]^{2+}(aq) + 4H_2O(l)$$

Substitution by chloride

It is also possible for chloride ions to substitute for water ligands. These reactions can involve a change in co-ordination number and shape as well as colour as chloride ions are bigger than water and ammonia. The overall equation for this reaction is:

$$[Cu(H_2O)_6]^{2+}(aq) + 4Cl^-(aq) \rightleftharpoons [CuCl_4]^{2-}(aq) + 6H_2O(l)$$

Cobalt(II)

When concentrated hydrochloric acid is added to aqueous cobalt(II) ions a tetrachlorocobaltate(II) complex is formed. If excess water is added, this reaction is reversed:

$$[Co(H_2O)_6]^{2+}(aq) + 4Cl^-(aq) \rightarrow [CoCl_4]^{2-}(aq) + 6H_2O(l)$$

$$[CoCl_4]^{2-}(aq) + 6H_2O(l) \rightarrow [Co(H_2O)_6]^{2+}(aq) + 4Cl^-(aq)$$

Copper(II)

When concentrated hydrochloric acid is added to aqueous copper(II) ions a tetrachlorocuprate(II) complex forms. This reaction is reversible:

$$[Cu(H_2O)_6]^{2+}(aq) + 4Cl^-(aq) \rightarrow [CuCl_4]^{2-}(aq) + 6H_2O(l)$$

$$[CuCl_4]^{2-}(aq) + 6H_2O(l) \rightarrow [Cu(H_2O)_6]^{2+}(aq) + 4Cl^-(aq)$$

Chelation

Complexes containing bidentate or mutidentate ligands are called **chelates**. The process of forming chelates is called **chelation**.

- Water, ammonia, and chloride ions are unidentate ligands; they can each donate one pair of electrons to the central metal ion.
- Bidentate ligands such as 1,2-diaminoethane and the ethanedioate ion can each donate two pairs of electrons.
- $EDTA^{4-}$ is a multidentate ligand; one ion can donate six pairs of electrons.

Bidentate ligands

There are two important bidentate ligands stated in the specification:

- 1,2-diaminoethane has two amino groups and so can donate two lone pairs.
- The ethanedioate ion has two carboxylate groups and can donate two lone pairs.

Up to three bidentate ligands may replace unidentate ligands in complexes. The ligands are replaced one by one in equilibria reactions with the position of equilibrium lying a long way to the right hand side. For example:

$$[Cu(H_2O)_6]^{2+} + 3H_2NCH_2CH_2NH_2 \rightleftharpoons [Cu(H_2NCH_2CH_2NH_2)_3]^{2+} + 6H_2O$$

$$[Cr(H_2O)_6]^{3+} + 3H_2NCH_2CH_2NH_2 \rightleftharpoons [Cr(H_2NCH_2CH_2NH_2)_3]^{3+} + 6H_2O$$

It is also possible for 1,2–diaminoethane to replace ammine ligands:

$$[Co(NH_3)_6]^{2+} + 3H_2NCH_2CH_2NH_2 \rightleftharpoons [Co(H_2NCH_2CH_2NH_2)_3]^{2+} + 6NH_3$$

The enthalpy and entropy changes in this reaction are interesting:

- The enthalpy change is almost zero because the same number and type of bonds are being broken.
- There is an increase in entropy so ΔS is positive as there are four molecules on the left hand side and seven on the right

  As a result the free energy change ΔG is negative so the reaction is feasible and stable chelates are formed.

HSW: EDTA⁴⁻

$EDTA^{4-}$ is a multidentate ligand which can donate six pairs of electrons to the metal ion. Remember that pairs of electrons can be donated by nitrogen atoms as well as from negatively charged ions. As a result $EDTA^{4-}$ can replace all six unidentate ligands in a complex:

$$[Cu(H_2O)_6]^{2+} + EDTA^{4-} \rightleftharpoons [Cu(EDTA)]^{2-} + 6H_2O$$

In this reaction

- The co-ordination number and oxidation state stay the same.
- The overall charge on the complex changes.
- The enthalpy change is almost zero.
- There is a positive entropy change.

  As a result the free energy change is negative.

$EDTA^{4-}$ is used in medicine to remove toxic heavy metal ions such as lead from the body. It is also used as a water softener as it forms chelates with the magnesium and calcium ions in hard water. It can be used as an anticoagulant to stop blood clotting.

Tetrahedral complexes

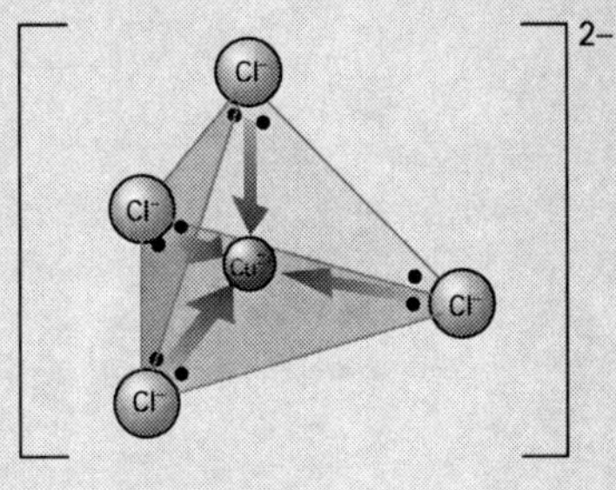

1,2-Diaminoethane and ethanedioate ion ligands

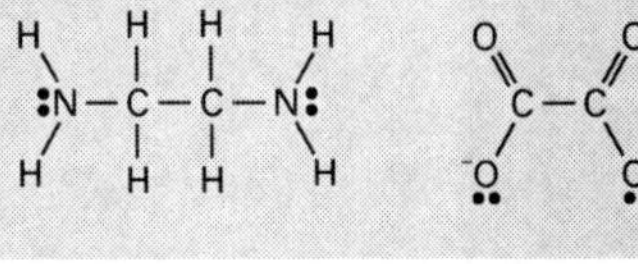

EDTA⁴⁻ ligand

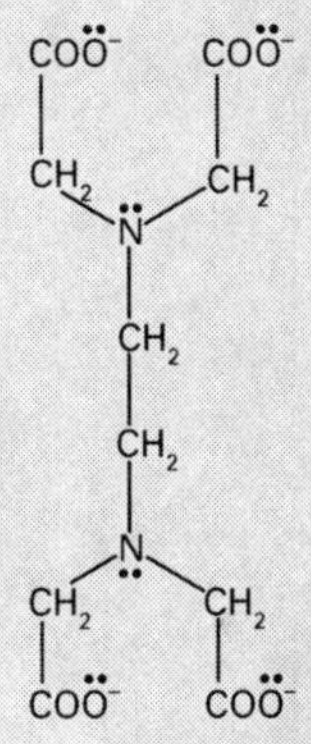

Questions

1. State the colour changes that occur when a small amount of aqueous ammonia is added to hexaaquachromium(II)(aq).
2. Explain why a green solution is often observed when concentrated hydrochloric acid is added to hexaaquacopper(II)(aq).
3. Explain why the replacement of ammonia ligands by 1,2-diaminoethane is favoured in free energy terms.

5.18 Reactions of cobalt and copper

Cobalt

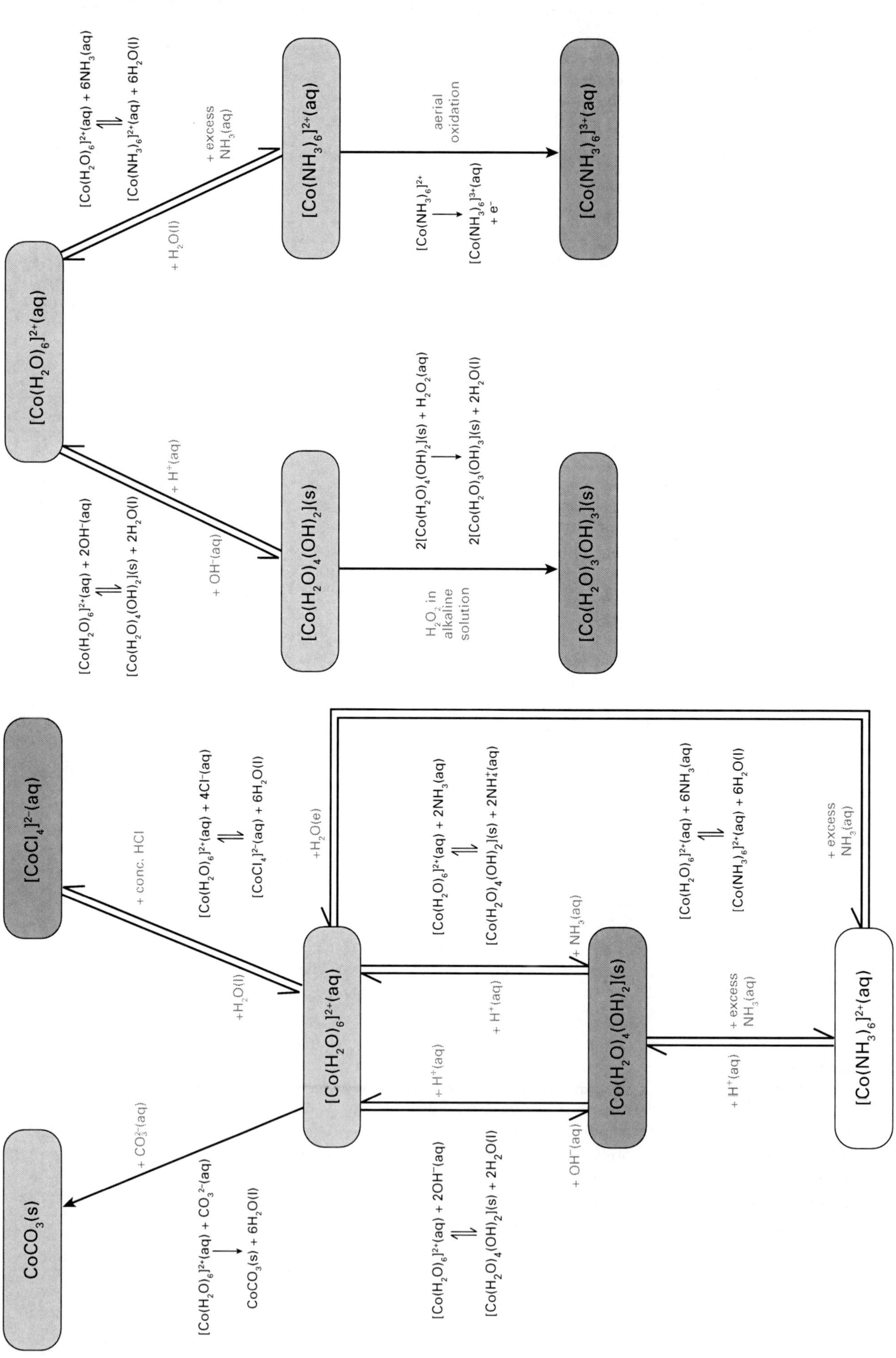

Copper

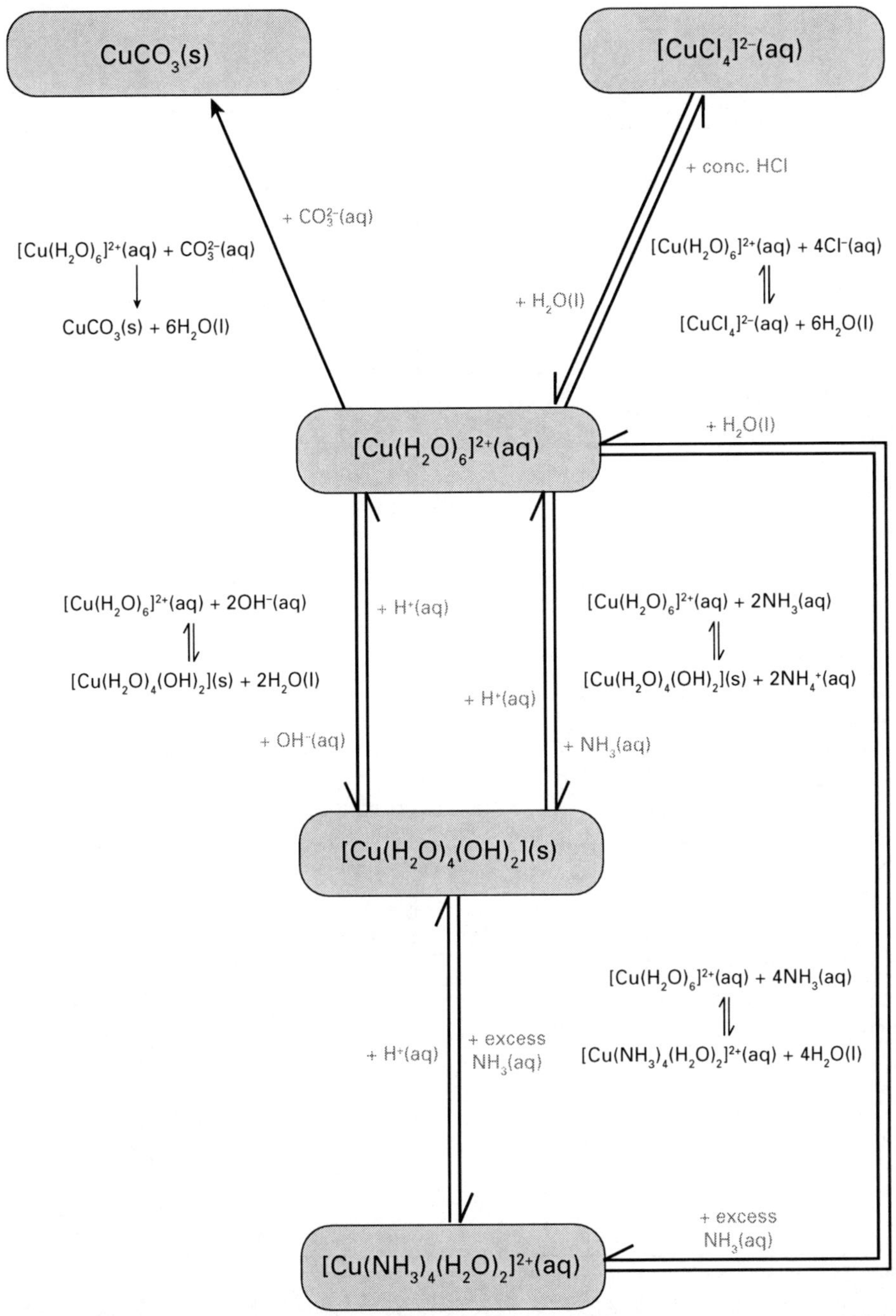

Summary of copper(II) colours

complex ion	colour
$[Cu(H_2O)_4(OH)_2](s)$	blue
$[Cu(H_2O)_6]^{2+}(aq)$	blue
$[Cu(NH_3)_4(H_2O)_2]^{2+}(aq)$	deep blue
$[CuCl_4]^{2-}(aq)$	olive-green

Summary of cobalt(II) and cobalt(III) colours

complex ion	colour
$[Co(H_2O)_4(OH)_2](s)$	blue-green
$[Co(H_2O)_6]^{2+}(aq)$	pink
$[Co(NH_3)_6]^{2+}(aq)$	straw-coloured
$[CoCl_4]^{2-}(aq)$	blue

Co(III) complex	colour
$[Co(H_2O)_3(OH)_3](s)$	dark brown
$[Co(NH_3)_6]^{3+}(aq)$	yellow

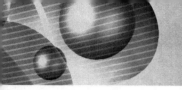

5.19 Reactions of aluminium, chromium, and iron

Aluminium and chromium

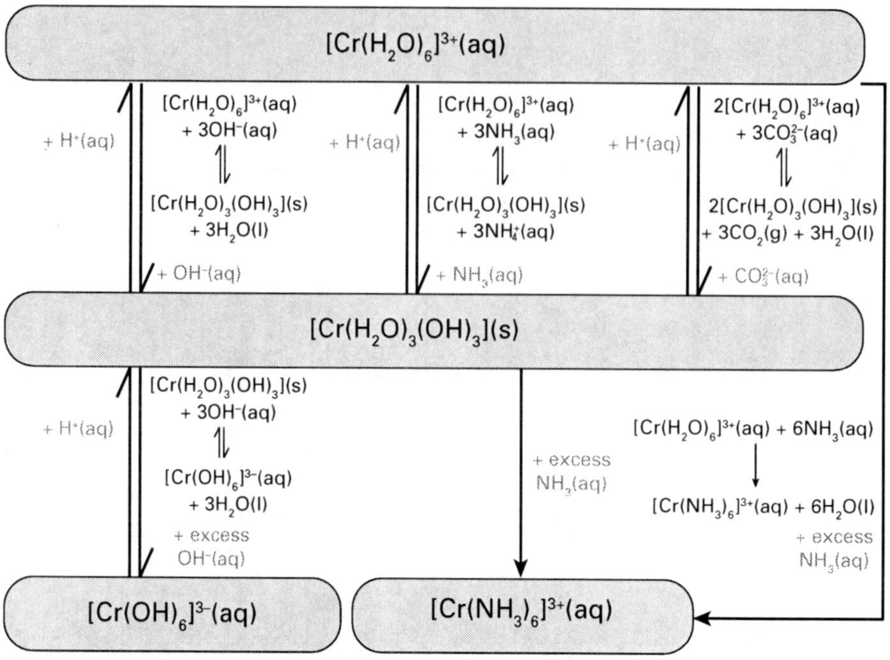

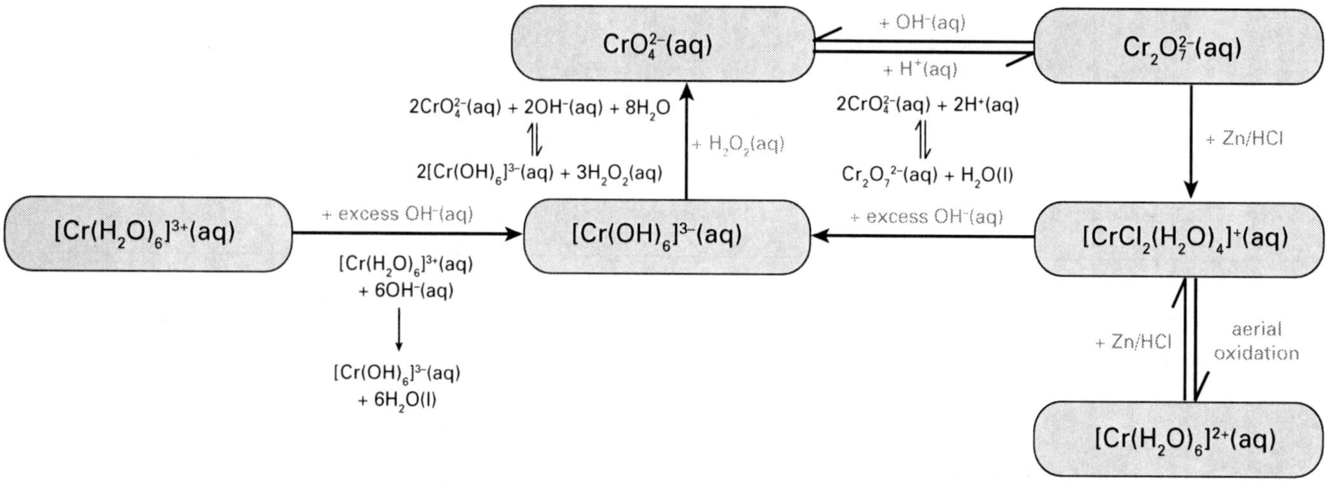

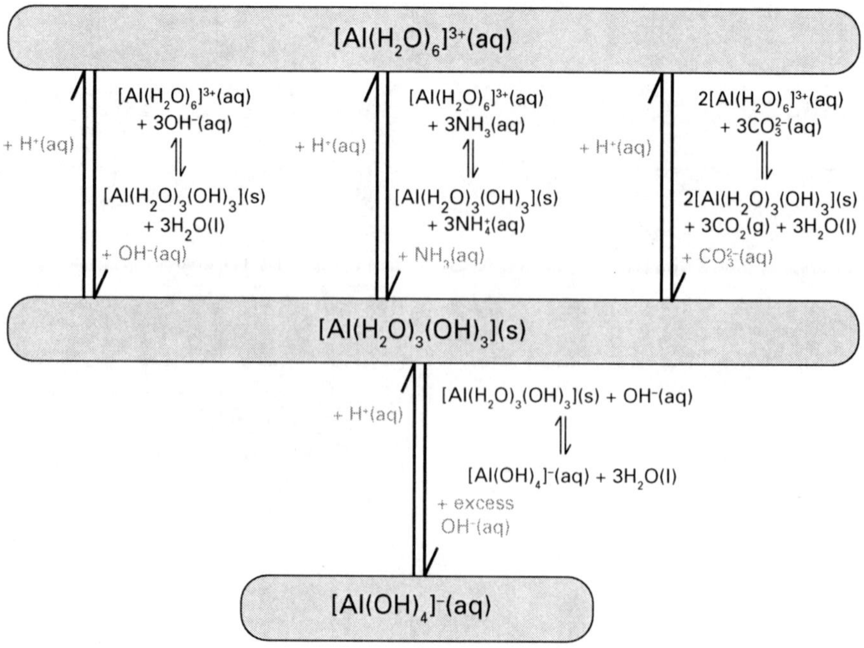

Iron

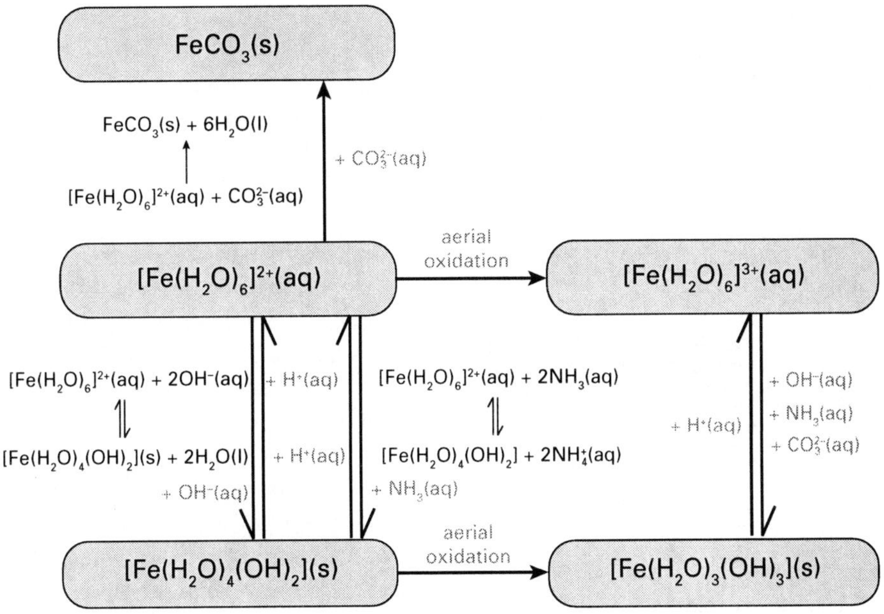

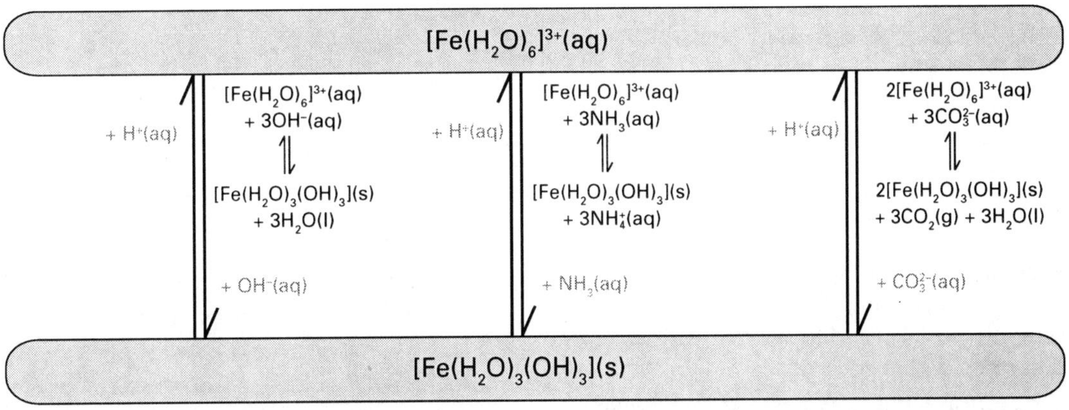

Summary of chromium colours

chromium(II) complex	colour
$[Cr(H_2O)_6]^{2+}(aq)$	blue

chromium(III) complex	colour
$[Cr(H_2O)_6]^{3+}(aq)$	ruby
$[Cr(H_2O)_3(OH)_3](s)$	grey-green
$[Cr(OH)_6]^{3-}(aq)$	dark green
$[Cr(NH_3)_6]^{3+}(aq)$	purple
$[CrCl_2(H_2O)_4]^+(aq)$	green

chromium(VI) complex	colour
$Cr_2O_7^{2-}(aq)$	orange
$CrO_4^{2-}(aq)$	yellow

Summary of iron colours

iron(II) complex	colour
$[Fe(H_2O)_6]^{2+}(aq)$	pale green
$[Fe(H_2O)_4(OH)_2](s)$	green
$FeCO_3(s)$	green

iron(III) complex	colour
$[Fe(H_2O)_6]^{3+}(aq)$	pale violet
$[Fe(H_2O)_3(OH)_3](s)$	brown

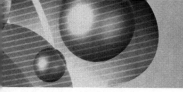

Further practice

Unit 4

1 Nitrogen monoxide, NO, reacts with hydrogen at 500 °C. The overall reaction can be summed up using this equation:

$$2NO(g) + 2H_2(g) \rightarrow N_2(g) + 2H_2O(g)$$

A series of experiments is carried out to find out about the rate of this reaction. These are the results of these experiments.

experiment	[NO] (mol dm^{-3})	[H_2] (mol dm^{-3})	Initial rate (mol dm^{-3} s^{-1})
A	0.2	0.2	3.0
B	0.2	0.4	6.0
C	0.6	0.4	54.0

a Work out the order of reaction with respect to each reactant. Show your reasoning.

b Write the rate equation for the reaction.

c What is the overall order of reaction?

d Calculate the rate constant, *k*, for this reaction. Include the units of *k*.

e Explain how the rate determining step is linked to the rate equation.

2 Ethane-1,2-diol can be used to manufacture a polymer.

a Give the displayed formula of a suitable compound that contains carbon, hydrogen, and oxygen only, which could be reacted with ethane-1,2-diol to produce a polymer.

b Give the repeat unit of the polymer that would be produced between ethane-1,2-diol and the molecule you have suggested .

c What type of polymer is this?

d Why would this polymer not be considered to be particularly hazardous to the environment?

e Polymers can be natural or synthetic.

i Name a synthetic polyester.

ii Name a natural polyamide.

3 Two straight chain alcohols labelled compound **A** and compound **B** have the molecular formula $C_4H_{10}O$. They both react with acidified potassium dichromate(VI).

a Compound **A** forms a carbonyl compound that does not react when it is warmed with Fehling's solution. Identify compound A.

b Compound **B** also forms a carbonyl compound. When this compound is warmed with Fehling's solution a brick-red precipitate of Cu_2O is formed. Identify compound **B**.

c One of the alcohols has two optical isomers. Draw the displayed formula of the alcohol that has two optical isomers. Label the chiral centre.

d Explain why this alcohol is optically active.

e Draw the two optical isomers of this alcohol.

4 **a** Define the ionic product of water, K_w

b Given that the ionic product of water, K_w, is given as 1.00×10^{-14} mol^2 dm^{-6}

i Calculate the pH of a 0.2 mol dm^{-3} solution of hydrochloric acid, HCl.

ii Calculate the pH of a 0.5 mol dm^{-3} solution of sodium hydroxide, NaOH.

c Calculate the pH of a solution in which the concentration of OH^- is 1000 times the concentration of H^+

d i Define a buffer solution

ii What can be added to a solution of ethanoic acid to make a buffer solution?

e A buffer solution is 0.01 mol dm^{-3} with respect to sodium ethanoate and 0.005 mol dm^{-3} with respect to ethanoic acid. Calculate the pH of the buffer solution given that for ethanoic acid K_a is 1.80×10^{-5} mol dm^{-3}.

5 Consider the following series of reactions:

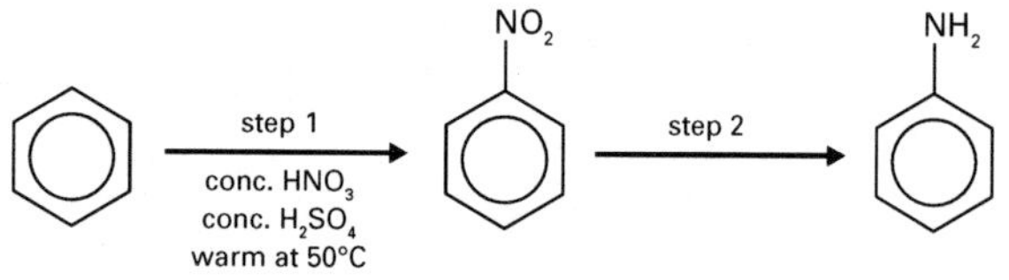

a What is meant by the term 'electrophile'?

b Identify the electrophile in step 1.

c Both nitric acid and sulfuric acid normally behave as acids. Consider the reaction below which happens in step 1. Use this to explain the term 'acid–base conjugate pair'.

$$H_2SO_4 + HNO_3 \rightarrow HSO_4^- + H_2NO_3^+$$

d Name the type of reaction mechanism involved in the nitration of benzene.

e i Name the organic product with the molecular formula $C_6H_5NH_2$. produced by step 2

ii Give the reagents and conditions used in step 2

6 Consider these two amino acids.

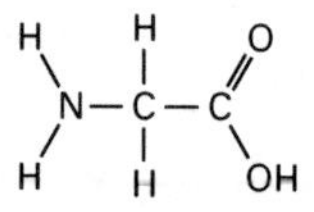

glycine

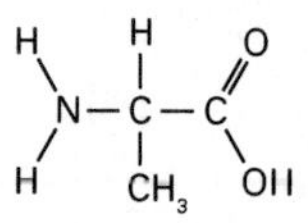

alanine

a Identify the functional groups in all amino acid.

b Explain why amino acids are amphoteric.

c i Which of these amino acids is a chiral molecule? Explain your answer.

ii How could you tell the two enantiomers of this amino acid apart?

d Near its isoelectric point an amino acid forms zwitterions. Draw the zwitterions of glycine.

e Two glycine molecules can join together to form a dipeptide.

i Draw the dipeptide produced.
ii Name this type of reaction.

f Amino acids can join together to form proteins. Describe what is meant by the primary structure of a protein.

7 The repeat units of two polymers are shown below.

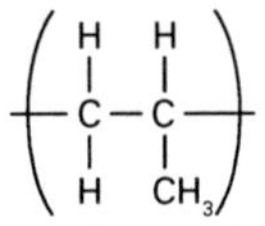

polymer **A**

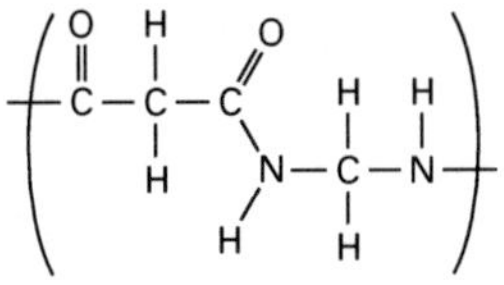

polymer **B**

a i Name and draw the structural formula of the monomer used to produce polymer **A**.
ii Name the type of polymerization used to produce polymer **A**.

b i Draw the structural formulae of the monomers used to produce polymer **B**.
ii Name the type of polymerization used to produce polymer **B**.

c Why is polymer **A** considered to be more hazardous to the environment than polymer **B**?

d Consider these two compounds.

i Draw the repeat unit of the polymer formed when these two compounds are reacted together.
ii What type of polymerization is this?
iii What type of polymer is formed?

8 The infrared spectrum of compound **A** is shown below.

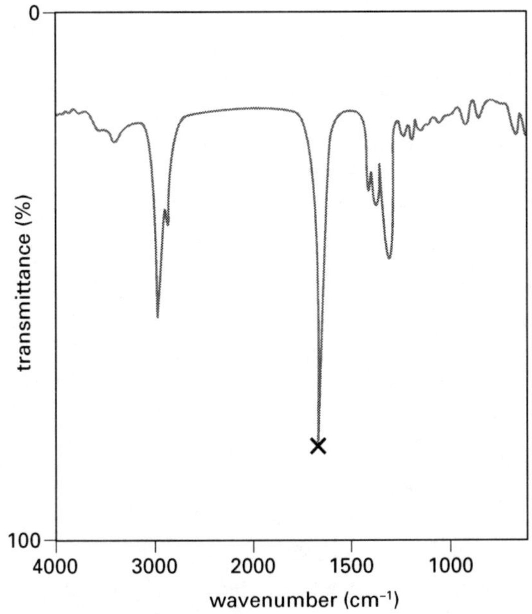

Consider the data on the type of bond and the wavenumber of radiation absorbed shown in the table to answer the following questions.

bond	wavenumber (cm^{-1})
O–H in carboxylic acids	2500–3300
O–H in alcohols	3230–3550
C–O	1000–1300
C=O	1680–1750
C–H	2850–3300
C–C	750–1100
C=C	1620–1680

Compound **A** contains carbon, hydrogen, and oxygen atoms.

a Identify the type of bond responsible for the absorption band marked X.

b Explain why compound **A** could be a ketone but could not be a secondary alcohol.

c Explain how the fingerprint region of an infrared spectrum could be used to identify compound **A**.

9 Kevlar is a useful polymer. It is made from the monomers shown below.

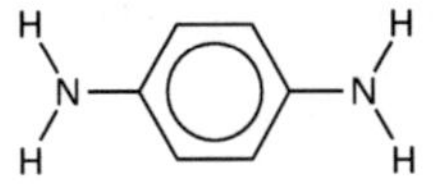

benzene-1,4-diamine

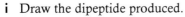

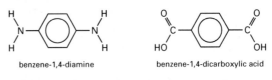

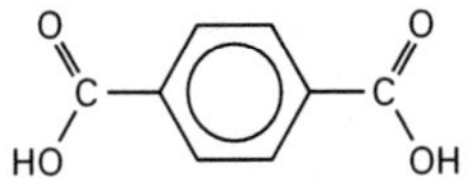

benzene-1,4-dicarboxylic acid

a Draw the repeat unit of Kevlar.

b Why is Kevlar a suitable material for making bullet-proof vests?

c What type of polymer is Kevlar?

d Why is Kevlar biodegradable?

e What is the name of the linkage between the monomers in Kevlar?

10 Ethanoic acid can be neutralized with sodium hydroxide solution to form sodium ethanoate:

$CH_3COOH(aq) + NaOH(aq) \rightarrow CH_3COONa(aq) + H_2O(l)$

25 cm^3 of 0.1 mol dm^{-3} ethanoic acid was placed into a conical flask. A 0.1 mol dm^{-3} solution of sodium hydroxide was placed in a burette and added to the flask until the pH remained constant.

a Sketch the pH curve for this titration.

The pH ranges for a selection of indicators are shown below.

indicator	pH range
methyl yellow	3.0–4.0
alizarin yellow	10.1–12.0
chlorophenol red	4.8–6.4

b Which of these indicators would you choose for this titration? Explain your answer.

c A buffer solution can made by dissolving sodium ethanoate in ethanoic acid. What is meant by the term 'buffer solution'.

Unit 5

1 a Define the term 'lattice formation enthalpy'.

b Complete the Born–Haber cycle shown below by writing the appropriate formulae with state symbols, on the dotted lines.

standard enthalpy change	substance referred to	enthalpy change (kJ mol^{-1})
enthalpy of atomization	chlorine	+121
enthalpy of atomization	magnesium	+150
enthalpy of formation	magnesium chloride	–642
first ionization enthalpy	magnesium	+736
second ionization enthalpy	magnesium	+1451
enthalpy of lattice formation	magnesium chloride	–2493

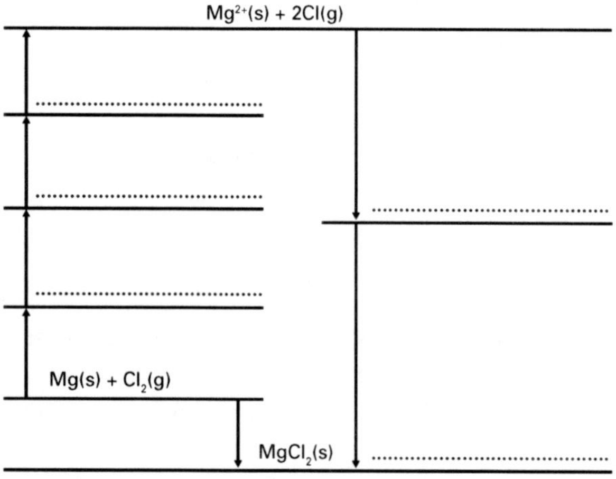

c Use the cycle and the values given to calculate the electron affinity of chlorine.

2 The graph shows how the entropy of a sample of water varies with temperature.

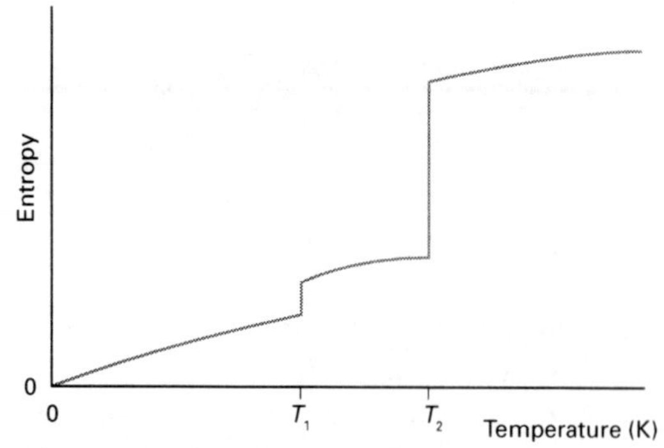

a State the change of state that occurs at temperature, T_1.

b The entropy change at temperature T_1 is much less than that at temperature T_2. Explain why.

c To convert 2.60 g of liquid water into steam at 373 K and 100 kPa takes 5.93 kJ of heat energy.

i Use this data to calculate the enthalpy change, ΔH, when 1.00 mol of liquid water forms 1.00 mol of steam at 373 K and 100 kPa.

ii State the relationship between free energy change, ΔG, enthalpy change, ΔH, and entropy change ΔS.

iii When liquid water is converted into steam at 373 K and 100 kPa the free energy change is 0 kJ mol^{-1}. Calculate the value of ΔS for the conversion of one mole of water into steam under these conditions.

3 Complete the table showing types of bonding.

oxide	structure	bonding
magnesium oxide		
aluminium oxide		
sulfur dioxide		

a Write balanced equations for the reactions of sodium oxide, aluminium oxide, and sulfur dioxide with water.

b Explain the difference in pH of each of these solutions.

4 Use the data in the table below, where appropriate, to answer the questions:

standard electrode potentials	$E^{\ominus}$ (V)
$Cl_2(aq) + 2e^- \rightarrow 2Cl^-(aq)$	+1.36
$Br_2(aq) + 2e^- \rightarrow 2Br^-(aq)$	+1.09
$Fe^{3+}(aq) + e^- \rightarrow Fe^{2+}(aq)$	+0.77
$I_2(aq) + 2e^- \rightarrow 2I^-(aq)$	+0.54

a Under suitable conditions, each of the reactions can be reversed.

i Identify the most powerful reducing agent in the table.

ii Identify the most powerful oxidizing agent in the table.

iii Explain why iodide ions can be used to reduce iron(III) ions but bromide ions cannot.

b The cell represented below was set up:

Pt | $Fe^{2+}(aq)$, $Fe^{3+}(aq)$ || $Cl_2(aq)$ | $Cl^-(aq)$ | Pt

i Calculate the e.m.f. of this cell.

ii Write a half-equation for the reaction occurring at the negative electrode.

iii What does the symbol || represent?

5 Fuel cells are able to convert the energy from a fuel such as hydrogen into electrical energy. In the hydrogen–oxygen fuel cell there are two flat electrodes, each coated on one side by a thin layer of platinum catalyst. A proton exchange membrane is placed between the two electrodes. Hydrogen gas flows to the anode and air to the cathode. Water vapour, which is the reaction product, is pushed out by the stream of air.

a Balance the half equations for the reactions which take place at the anode and the cathode:

Anode: $H_2(g) \rightarrow 2H^+(aq)$

Cathode $4H^+(aq) + O_2(g) \rightarrow 2H_2O(g)$

b State with a reason which of these reactions is oxidation.

c Combine the two half-equations to show the overall reaction which occurs in the fuel cell.

d State one advantage and one disadvantage of using hydrogen in a fuel cell.

6 a The reaction of the manganate(VII) ion with iron(II) ions can be used to determine the mass of iron(II) sulfate present in an iron tablet. The half-equations for the reduction of manganate(VII) and the oxidation of iron(II) are:

$$MnO_4^-(aq) + 8H^+(aq) + 5e^- \rightarrow Mn^{2+}(aq) + 4H_2O(l)$$

$$Fe^{2+}(aq) \rightarrow Fe^{3+}(aq) + e^-$$

i Combine these two half-equations to give an overall equation for the reaction of manganate(VII) with iron(II).

ii State the colour changes of the manganate(VII) ion in this reaction.

b An iron pill is crushed in a pestle and mortar then dissolved in 100 cm^3 of sulfuric acid. 25.0 cm^3 of this solution is titrated against 0.1 mol dm^{-3} potassium manganate(VII) solution. 23.40 cm^3 of this solution is needed to fully oxidize the iron(II) ions.

i Calculate the number of moles of manganate(VII) ions used in the titration.

ii Calculate the number of moles of iron(II) ions oxidized by the manganate(VII).

iii Calculate the mass of iron present in the iron pill.

7 The following scheme shows some reactions of copper compounds in aqueous solution.

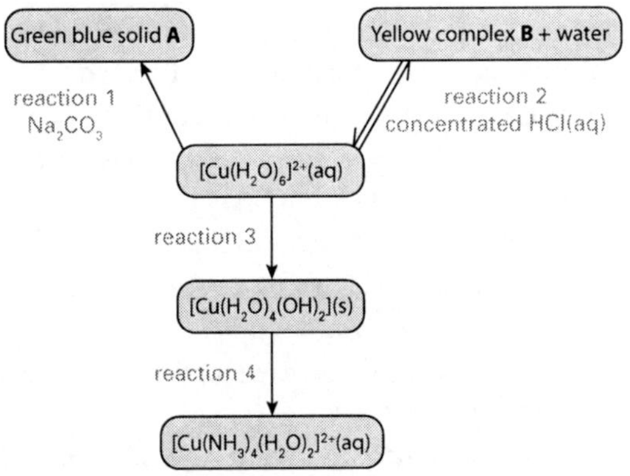

a Identify the green-blue solid **A** formed in reaction 1. Write an equation for the reaction.

b Identify the yellow complex **B** formed in reaction 2. Write an equation for the reaction.

c Draw the complex formed in reaction 2.

d i Identify the reagent needed for reaction 3.

ii Write an equation for the reaction.

e i Identify the reagent needed for reaction 4.

ii Name the type of reaction taking place.

8 Ammonia is able to act as both a Brønsted–Lowry and a Lewis base in its reactions with chromium.

a Draw the ammonia molecule, labelling clearly the N–H bond angle.

b Define the term Brønsted–Lowry base.

c Define the term Lewis base.

d Complete the two equations given below:

Reaction 1:

$[Cr(H_2O)_6]^{3+}(aq) + \ldots\ldots \rightleftharpoons [Cr(H_2O)_3(OH)_3](s) + 3NH_4^+(aq)$

Reaction 2:

$[Cr(H_2O)_6]^{3+}(aq) + \ldots\ldots\ldots \rightarrow [Cr(NH_3)_6]^{3+}(aq) + \ldots\ldots\ldots$

e State the reaction in which ammonia is acting as a Lewis base.

9 a Both aluminium and iron form hexaaqua complexes in aqueous solution.

i State the formula of the complex formed by aluminium.

ii State the formulae of the two possible complexes formed by iron.

iii Which of the two iron complexes has the lower pH? Explain your answer.

b When aqueous sodium hydroxide is added to aqueous aluminium ions a white precipitate of aluminium hydroxide forms. Write a balanced equation for this reaction.

c Aluminium hydroxide is amphoteric. Explain the meaning of the term amphoteric using relevant reactions of aluminium hydroxide in your answer.

10 Transition metals and their compounds are often used as catalysts.

a Define the term catalyst.

b Sketch an energy level diagram to show the action of a catalyst.

c Explain the term 'homogeneous catalyst'.

d Use the equations below to identify the catalyst for the decomposition of hydrogen peroxide to form water and oxygen.

Step 1:

$H_2O_2(aq) + Br_2(aq) \rightarrow 2H^+(aq) + 2Br^-(aq) + O_2(g)$

Step 2:

$H_2O_2(aq) + 2H^+(aq) + 2Br^-(aq) \rightarrow 2H_2O(l) + Br_2(aq)$

e Name the catalyst used for the Haber process.

f Explain what is meant by a 'catalyst support' and explain why a catalyst support is used in the Haber process.

Synoptic questions

Your examination will include some synoptic questions: questions that involve more than one section of the specification. These are a selection of synoptic questions based on Units 4 and 5 to give you practice in answering questions of this style. There is an example of a synoptic question/answer on p 89.

Unit 4

1 Compound **A** is a straight chain organic acid. A sample of the acid has a percentage composition by mass of C, 59%; H, 31%; O, 10%.

a Calculate the empirical formula of the acid.

b The M_r of the acid is found to be 102. Calculate the molecular formula of the acid.

c Suggest the systematic name of this compound.

d Suggest how you could confirm that this organic compound is a carboxylic acid.

e Ethanoic acid is a weak Brønsted–Lowry acid. What is meant by the term 'weak acid'?

f Given that K_a for ethanoic acid is 1.78×10^{-5} mol dm^{-3} calculate the pH of a 0.2 mol dm^{-3} solution of ethanoic acid.

2 Compound **A** is a primary alcohol. It has the percentage composition by mass of C, 60%; H, 13%; O, 27%.

a Calculate the empirical formula of compound **A**.

b If M_r of compound **A** is 60, give the molecular formula and name of compound **A**.

c Consider the series of reactions below.

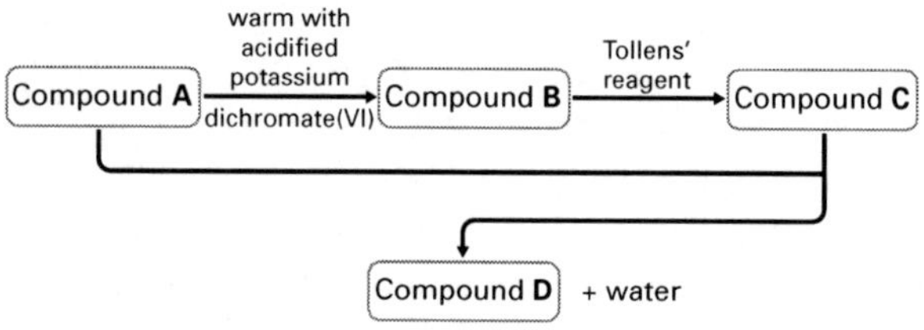

Compound **A** is warmed with acidified potassium dichromate(VI). One of the products of the reaction is an organic compound labelled **B**.

i What would you SEE happen?

ii What type of reaction is this?

iii Identify compound **B**.

d Identify organic compound **C**.

e Compound **A** is mixed with compound **C**. An acid catalyst is added and the mixture is warmed. An organic compound labelled **D** and water are formed. Identify the organic product formed in this reaction.

3 Consider the series of reactions shown below.

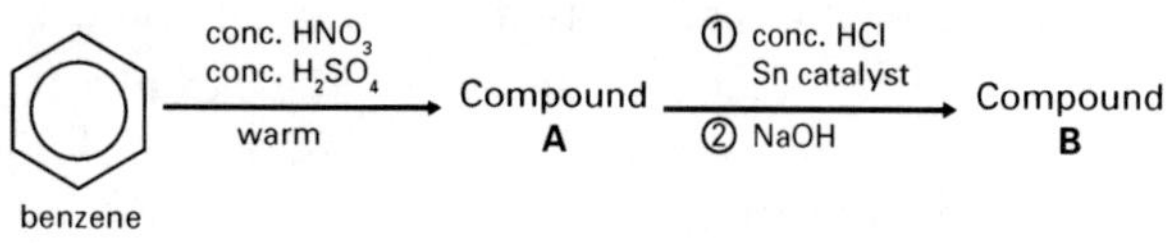

a Identify compounds A and compound B.

b Explain why this series of reactions is not normally carried out in schools.

c Name the type of reaction that takes place in step 1 and step 2.

d i Identify the electrophile in step 1

ii Give the mechanism for the reaction which takes place between the electrophile and benzene in step 1.

4 Give the reagents and observations you would use to identify the following functional groups.

a

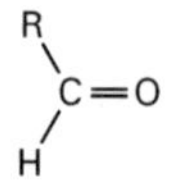

b

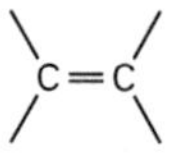

Butene exists as two positional isomers; but-1-ene and but-2-ene.

c But-2-ene exists as two geometric isomers. Draw and label each of these geometric isomers.

d But-1-ene does not have geometric isomers. Explain why.

5 Consider the diagram below.

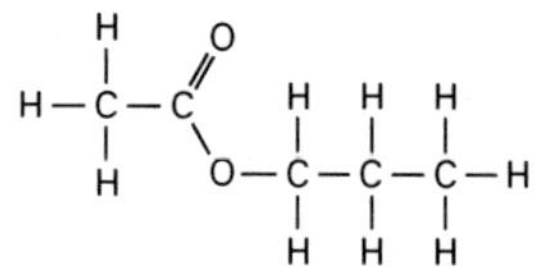

compound **A**

a i What type of compound is compound **A**?

ii Name compound **A**.

b Compound **A** can be hydrolysed by refluxing with dilute sulfuric acid. Draw and name the compounds produced by the hydrolysis of compound **A**.

c Compound **A** can also be hydrolysed by refluxing with sodium hydroxide solution. Give one difference between the products produced in this reaction and the products made in reaction **b**.

d A sample of ethanol is analysed using ^{1}H n.m.r integrated spectra. Suggest the features that would be seen in the spectrum of ethanol.

6 Nitrogen and hydrogen can react to produce ammonia. An iron catalyst can be used.

$N_2(g) + 3H_2(g) \rightleftharpoons 2NH_3(g) \quad \Delta H = -92$ kJ mol^{-1}

a What is the significance of the sign of the enthalpy change?

b How does the iron catalyst affect:

i the time taken to reach equilibrium?

ii the yield of ammonia?

c The temperature of the reaction is increased. How does this affect:

i the time taken to reach equilibrium?

ii the yield of ammonia?

d The pressure of the reaction is increased. How does this affect the yield of ammonia?

e Write an expression for K_c for this equilibrium.

f The reaction is carried out at 500 °C. At this temperature $K_c = 8.00 \times 10^{-2}$ mol^{-2} dm^6.
At equilibrium the concentration of H_2 = 3.0 mol dm^{-3} and the concentration of N_2 = 1.8 mol dm^{-3}. Calculate the equilibrium concentration of NH_3

7 Ammonium nitrate, NH_4NO_3, decomposes to form nitrous oxide, N_2O, and water, H_2O.

$NH_4NO_3(s) \rightarrow N_2O(g) + 2H_2O(l)$

Assume 1 mole of a gas occupies 24 dm^3.

a 50 g of ammonium nitrate is heated. Calculate:

i the mass of water produced

ii the volume of nitrous oxide produced

b Calculate the oxidation number of each nitrogen atom in ammonium nitrate, $NH_4^+NO_3^-$

c On heating, nitrous oxide decomposes to form nitrogen and oxygen.

$2N_2O(g) \rightarrow 2N_2(g) + O_2(g)$

Use the graph below to show that this reaction is first order with respect to nitrous oxide.

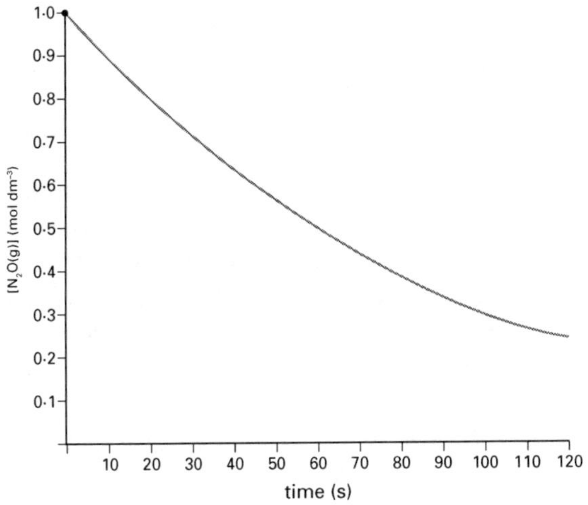

d Write a rate equation for this reaction.

8 Look at the repeat units of the polymers below

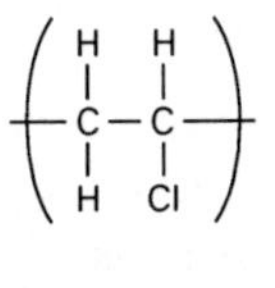

polymer **A**

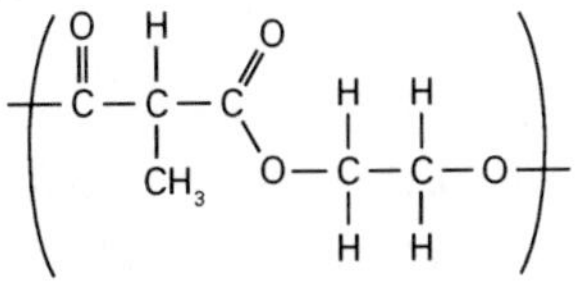

polymer **B**

a i Name the monomer used to make polymer **A**.

ii Name the type of reaction used to make polymer **A**.

b i Name the monomers used to make polymer **B**.

ii Name the type of reaction used to make polymer **B**.

c What type of polymer is **B**? Why would polymer **B** not be suitable for use with concentrated alkalis?

9 Haloalkanes may be identified by warming with aqueous sodium hydroxide, neutralizing the resulting mixtures with dilute nitric acid, and finally adding an aqueous solution of silver nitrate.

a Complete the table to show the colour of the precipitate formed and the identity of the precipitate.

haloalkane	colour of precipitate	identity of the precipitate
chlorooalkane		
bromoalkane		
iodoalkane		

b Explain how you could use ammonia to confirm that a haloalkane was a bromoalkane.

c Consider the mass spectrum of 1-chloropropane shown below.

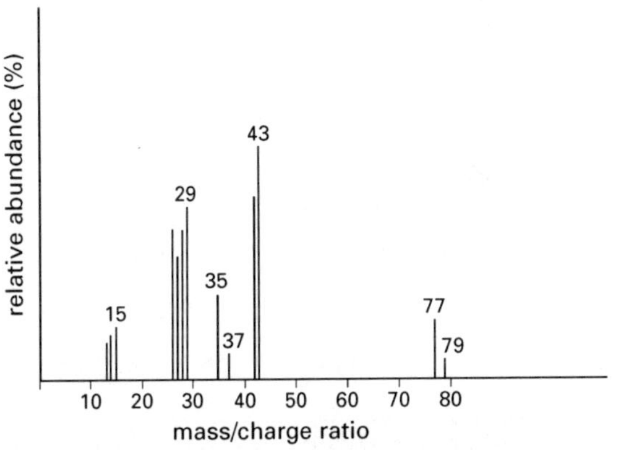

i Identify the ions responsible for the peaks at 15, 29, and 43.

ii Why are there peaks at 77 and at 79?

10 Propanoic acid reacts with hydrogen to produce propanal.

$CH_3CH_2COOH + H_2 \rightarrow CH_3CH_2CHO + H_2O$

a What type of reaction is this?

b Calculate the enthalpy change for the reaction between propanoic acid and hydrogen, given these enthalpy changes of combustion:

	enthalpy of combustion (kJ mol^{-1})
propanoic acid	–1527
hydrogen	–286
propanal	–1821

c Write an expression for the acid dissociation constant, K_a, of propanoic acid.

d i K_a for propanoic acid is 1.30×10^{-5} mol dm^{-3}. Calculate the pH of a 0.2 mol dm^{-3} solution of propanoic acid.

ii What assumptions have you made in your calculation?

Unit 5

1 Most elements can be extracted from one of their naturally occurring compounds by one of the following methods:

- reduction of the oxide with carbon
- displacement of the element from its compound by a more reactive element

a Define the term 'standard enthalpy of formation'.

b Explain why the standard enthalpies of formation of Al(s) and C(s) are zero.

c Use the information given to show that the carbon reduction process for the extraction of aluminium is not feasible at 350 K.

$Al_2O_3(s) + 3C(s) \rightarrow 2Al(s) + 3CO(g)$

substance	$\Delta H_f^{\ominus}$ (kJ mol–1)	$S^{\ominus}$ (J K^{-1} mol^{-1})
$Al_2O_3(s)$	–1669	51
Al(s)	0	28
CO(g)	–111	198
C(s)	0	6

d i State the method used to reduce aluminium oxide in industry.

ii Give the essential conditions for this industrial process.

2 a Complete the table showing types of bonding

substance	structure	bonds broken on melting
sodium		
silicon		
phosphorus		

b Sodium and chlorine are added separately to water. Write equations for any reactions that take place and state the approximate pH of any solutions formed.

c Using suitable equations for the reactions of aluminium compounds, explain what is meant by the term 'amphoteric'.

3 a Chromium(III) can be oxidized to chromium(VI) in alkaline solution using hydrogen peroxide, H_2O_2. This reaction occurs in three steps. In the first step aqueous sodium hydroxide solution is added to Cr(VI) forming $[Cr(H_2O)_3(OH)_3](s)$. This then reacts with excess sodium hydroxide solution forming a dark green solution. In the third step this solution is warmed with hydrogen peroxide forming a yellow solution of $CrO_4^{2-}(aq)$.

i Draw a diagram to show the shape of the hydrogen peroxide molecule and label the H–O–O bond angle.

ii Write out in full the electron configuration of the Cr(III) ion.

iii Write a balanced equation for the reaction occurring in step 1.

iv Give the formula of the chromium-containing species formed in step 2.

v Write a balanced equation for step 3.

b Another ion containing chromium(VI) is the dichromate ion, $Cr_2O_7^{2-}(aq)$. This can be used in redox titrations to find the concentrations of solutions of iron(II) ions. In such a titration 10.40 cm^3 of $Cr_2O_7^{2-}(aq)$ was used to oxidize exactly 25.0 cm^3 of 0.05 mol dm^{-3} $Fe^{2+}(aq)$.

i Balance the equation below for the oxidation of iron(II) ions with acidified dichromate ions:

$Cr_2O_7^{2-}(aq) + H^+(aq) + Fe^{2+}(aq) \rightarrow Cr^{3+}(aq) + H_2O(l) + Fe^{3+}(aq)$

ii Calculate the concentration in mol dm^{-3} of the $Cr_2O_7^{2-}(aq)$ ions.

iii What volume of 0.1 mol dm^{-3} sulfuric acid would be needed to acidify the dichromate solution?

4 The bidentate ligand ethane-1,2-diamine can be synthesized in two stages starting from ethene. This is shown in the reaction scheme below:

$$H_2C{=}CH_2 \xrightarrow{\text{reaction 1}} BrCH_2CH_2Br \xrightarrow{\text{reaction 2}} H_2NCH_2CH_2NH_2$$

a i Suggest a suitable reagent for reaction 1.

ii Name and outline the mechanism for reaction 1.

b i Write a full balanced equation for reaction 2.

ii Calculate the atom economy for this reaction and compare it with that for reaction 1.

c Ethane-1,2-diamine is able to act as a Lewis base. Explain what is meant by a 'Lewis base'.

d i Explain what is meant by the term 'bidentate ligand'.

ii Why is ethane-1,2-diamine able to act as a bidentate ligand?

e Draw the structure of the complex formed when aqueous copper(II) ions react with an excess of ethane-1,2-diamine.

f The formation of the complex in part e is said to be feasible. Explain this statement.

5 The transition metals have a number of characteristic properties which give them many uses in industry.

a List two of the characteristic properties of the transition metals.

b Explain the difference between heterogeneous and homogeneous catalysis. In your answer include equations for:

i the action of Fe^{2+} ions as a homogeneous catalyst in the reaction between I^- and $S_2O_8^{2-}$ ions

ii the action of vanadium(V) oxide as the catalyst in the contact process

6 Use the electrode potential data to help you answer the question below.

	Eo (V)
$Fe^{3+}(aq) + e^- \rightarrow Fe^{2+}(aq)$	+0.77
$Fe^{2+}(aq) + 2e^- \rightarrow Fe(s)$	–0.44
$Zn^{2+}(aq) + 2e^- \rightarrow Zn(s)$	–0.76
$Cl_2(g) + 2e^- \rightarrow 2Cl^-(aq)$	+1.36
$O_2(g) + 2H_2O(l) + 4e^- \rightarrow 4OH^-(aq)$	+0.40

Each of the reactions above can be reversed under suitable conditions

a Explain why iron is described as a transition metal and a d block element while zinc is only described as a d block element.

b Combine suitable half-equations to show the oxidation of iron by air in the presence of water.

c **i** What is the appearance of the product of the reaction in part **b**?

ii How does its appearance change if it is left in air? State the formula of the product formed.

d **i** Explain the term 'reducing agent'.

ii Why are iron(III) ions able to oxidize zinc ions but not able to oxidize chloride ions?

Synoptic worked example

2-Promo-methylpropane reacts with hydroxide ions to produce methylpropan-2-ol:

$$C_4H_9Br + OH^- \rightarrow C_4H_9OH + Br^-$$

The table below shows the results from a series of experiments.

experiment	$[C_4H_9Br]$ (mol dm^{-3})	$[OH^-]$ (mol dm^{-3})	Initial rate (mol dm^{-3} s^{-1})
A	0.1	0.1	0.056
B	0.1	0.2	0.056
C	0.2	0.1	0.112
D	0.2	0.2	0.112

What is the order of reaction with respect to 2-bromo-methylpropane and with respect to hydroxide ions? Explain your reasoning. (2)

The order of reaction with respect to 2-bromo-methylpropane is 1. When the concentration of 2-bromo-methylpropane is doubled the rate of reaction is doubled.

The order of reaction with respect to hydroxide ions is 0. When the concentration of hydroxide ions is doubled the rate of reaction is unchanged.

The student has answered the question and carefully explained their reasoning. They score the marks.

What is the overall order of reaction? (1)

1 + 0 = 1

An easy mark. The overall order of reaction is the sum of the individual orders of reaction.

What is the rate equation for the reaction? (1)

Rate = $k[C_4H_9Br]$

Well done. Remember to use square brackets to indicate that concentration is being measured in mol dm^{-3}

Calculate the value of *k*, including units. (1)

Rate = $k[C_4H_9Br]$
$0.056 = k \times 0.1$
$k = 0.056/0.1$
$k = 0.56$

Although the student has calculated the numerical part of the answer they have forgotten to include the units (s^{-1}) so do not get the mark.

Consider the reaction mechanisms shown below. Which of these possible mechanisms is consistent with the kinetic evidence of how 2-bromo-methylpropane reacts with hydroxide ions? Explain your reasoning (1)

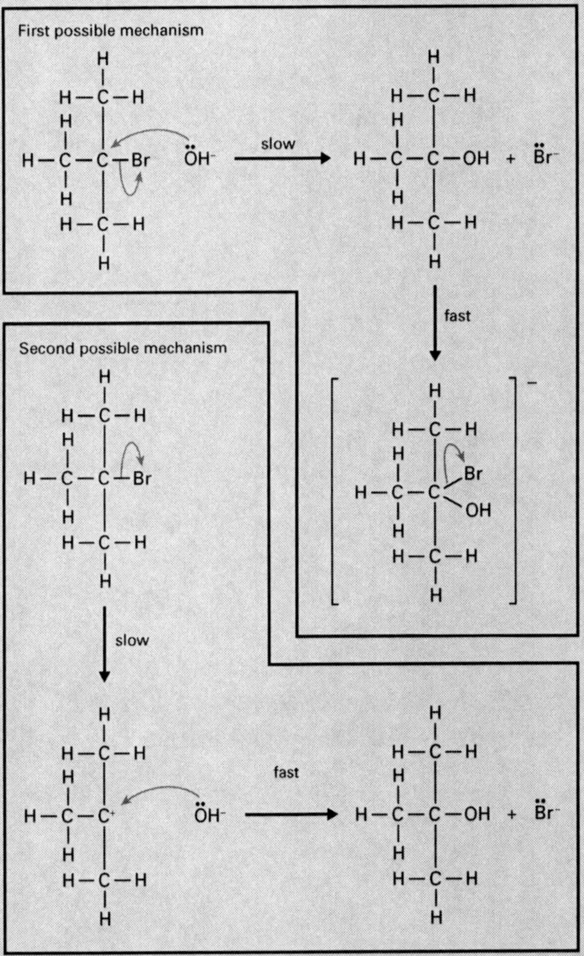

The rate equation tells us that C_4H_9Br must be in the rate determining step but that OH^- is not.

This means that the first mechanism is not correct because both species are shown as being in the rate determining step.

The second mechanism is possible. It shows that only C_4H_9Br is in the rate determining step.

The student has identified the second mechanism as being consistent with the evidence. They have explained their answer and score the mark.

Answers

The answers are in three parts:
- answers to text spread calculations
- answers to the further pratice
- answers to the synoptic questions

Answers to text spread calculations
Unit 4

4.01

1 4

2 mol–1 dm3 s–1

3 rate = k[A]

4.03

3 $K_c = \frac{[C]^2}{[A]^2\,[B]^3}$

4.04

1 a $K_c = \frac{[C]}{[A]\,[B]^2}$

b A = 0.06 mol dm^{-3}, B = 0.4 mol dm^{-3}, C = 0.08 mol dm^{-3} **c** 8.33 mol^{-2} dm^6

2 0.4 moles of Y and 0.2 moles of Z

3 0.01 moles of A, 0.08 moles of B, and 0.02 moles of C

4.05

2 a 1.00; **b** 0.30; **c** 0.22

Unit 5

5.01

1 $\Delta H_r^{\ominus}$ = –2046 kJ mol^{-1}

2 $\Delta H_r^{\ominus}$ = +1388 kJ mol^{-1}

5.02

1 b $\Delta H_{soln}^{\ominus}$ = –35 kJ mol^{-1}

5.03

1 b Lattice dissociation enthalpy = –723 kJ mol^{-1}
Negative value means that MgCl is unstable.

5.04

3 T = 1079 K

5.06

1 a +7 **b** +2 **c** +5 **d** –3

5.08

2 a $E^{\ominus}_{cell}$ = +0.32 V **b** $E^{\ominus}_{cell}$ = +0.55 V
c $E^{\ominus}_{cell}$ = +0.74 V

5.13

1 potassium dichromate(VI) concentration = 8.8 × 10^{-3} mol dm^{-3}

2 volume of sulfuric acid = 0.419 dm^3 = 41.9 cm^3

Answers to further practice
Unit 4

1 a Second order with respect to NO. When the concentration is tripled the rate increases by 3^2 or nine times. First order with respect to H_2. When the concentration is doubled the rate doubles.
b Rate = k$[NO]^2[H_2]$ **c** Overall order of reaction = 2 + 1 = 3 **d** 375 mol^{-2} dm^6 s^{-1} **e** The species in the rate equation will be involved in the rate determining step

2 a A dicarboxylic acid, for example

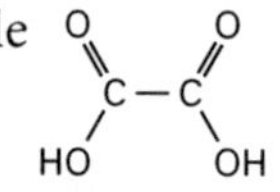

b Using the dicarboxylic acid suggested

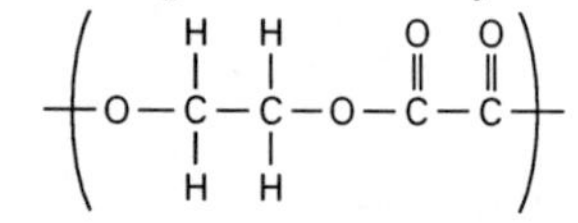

c a polyester **d** It is biodegradable/will be hydrolysed/will react with acids or bases.
e i Terylene/PET **ii** polypeptide/protein

3 a butan-2-ol **b** butan-1-ol

c

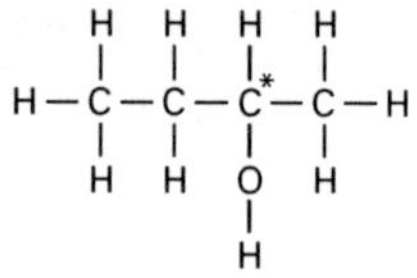

d The second carbon has four different groups bonded to it.

e

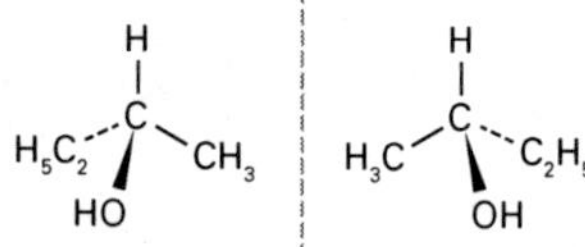

4 a $K_w = [H^+][OH^-] = 1.00 \times 10^{-14}$ mol^2 dm^{-6}
b i 0.70 **ii** 13.70 **c** 8.50 **d i** A buffer resists changes to the pH when small amounts of acid or base are added. **ii** sodium/sodium ethanoate **e** 5.05

5 a An electrophile is a species which can accept a lone pair of electrons. **b** Nitronium ion, NO_2^+
c The species formed when an acid loses an H^+ ion is the acid's conjugate base. HSO_4^- is the conjugate base of H_2SO_4 and $H_2NO_3^+$ is the conjugate acid of HNO_3. The pairs are linked by the gain or loss of an H^+ ion. **d** Electrophilic substitution/nitration
e i phenylamine **ii** First, concentrated HCl and tin and reflux, then NaOH(aq)

6 a carboxyl, COOH, and amino, NH_2
b They react with acids and with bases.
c i Alanine has a chiral molecule – it has a carbon atom bonded to four different groups. **ii** Use plane polarized light. One enantiomer rotates the plane of polarization clockwise while the other enantiomer turns it anticlockwise.

d

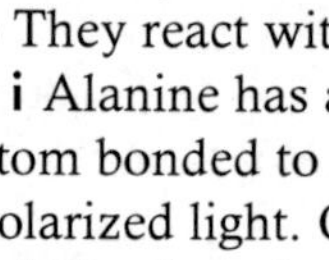

e i

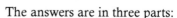

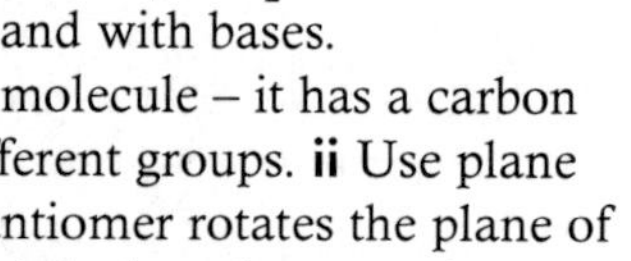

ii condensation polymerization
f the order of the amino acids in the protein

7 a i 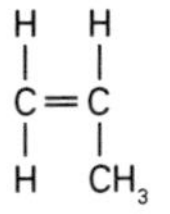

ii addition polymerization

b i 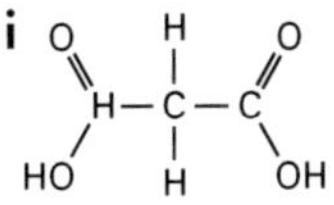

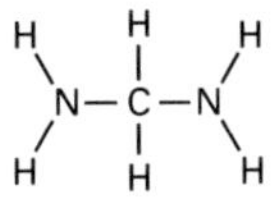

ii condensation polymerization **c** Condensation polymers can be hydrolysed so they break down more rapidly than addition polymers.

d i 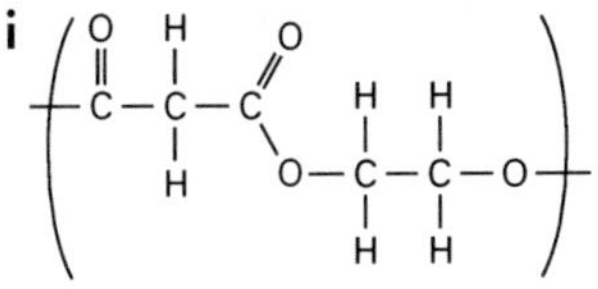

ii condensation polymerization **iii** polyester

8 a C=O **b** It does have a peak at 1700 cm^{-1} and does not have a peak at 3230–3550 cm^{-1} **c** The fingerprint of compound A can be compared with a library of fingerprints of known compounds.

9 a 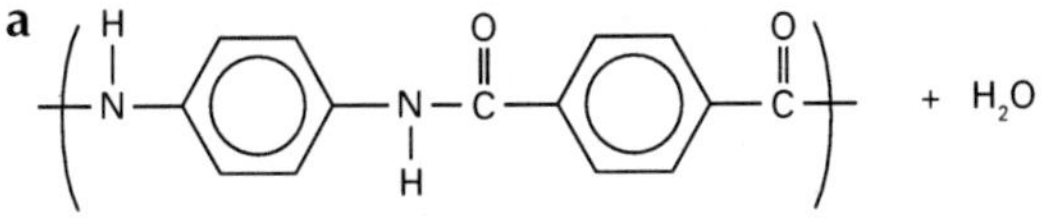

Kevlar

b It is strong and lightweight. **c** polyamide **d** It can be hydrolysed. **e** amide link

10 a Ethanoic acid will have a pH of around 3 and sodium hydroxide will have a pH of around 13. So the graph starts at 3 and ends at 13. The graph is vertical between pH 5 and pH 11.The equivalence point (when neither acid nor alkali is present in excess) is at pH 9. **b** Chlorophenol red because its pH range coincides with the steep rise of the pH curve of the graph. **c** A buffer solution resists a change in pH when small amounts of acid or base are added to it.

Unit 5

1 a Lattice formation enthalpy is the enthalpy change when one mole of an ionic solid is formed from its gaseous ions.

b 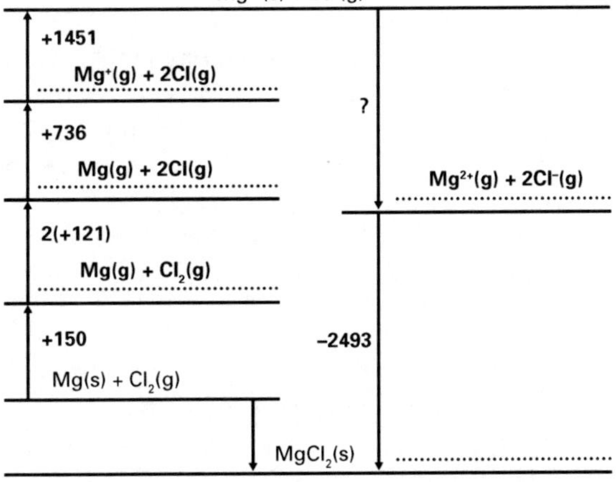

c Electron affinity of chlorine = –364 kJ mol^{-1} (Check carefully if you got –728 kJ mol^{-1})

2 a T_1 is melting **b** At temperature T_1 liquid water is being formed; at T_2 steam is being formed. There is a slightly greater degree of disorder in liquid water compared with solid water but a very much greater degree of disorder in gaseous water. **c i** $\Delta H = 41.1$ kJ **ii** $\Delta G = \Delta H - T\Delta S$ **iii** $\Delta S = 0.11$ kJ mol^{-1}

3

oxide	structure	bonding
magnesium oxide	giant	ionic
aluminium oxide	giant	ionic
sulfur dioxide	giant	covalent

a $Na_2O(s) + H_2O(l) \rightarrow 2NaOH(aq)$
Aluminium oxide no reaction
$SO_2(g) + H_2O(l) \rightleftharpoons H_2SO_3(aq)$
b Sodium oxide is basic so forms a solution of pH 14; sulfur dioxide is an acidic oxide so forms a solution of pH 1.

4 a i iodide ions, I^- **ii** chlorine solution **iii** The potential for the reduction of the Fe^{3+} ions is more positive than that for iodine and more negative than that of bromine. This means that Br_2 is a stronger oxidizing agent than Fe^{3+} so they will not react. **b i** 0.59 V **ii** $Fe^{2+}(aq) \rightarrow Fe^{3+}(aq) + e^-$ **iii** the salt bridge

5 a Anode: $H_2(g) \rightarrow 2H^+(aq) + 2e^-$
Cathode: $4H^+(aq) + 4e^- + O_2(g) \rightarrow 2H_2O(g)$
b The reaction which takes place at the anode is oxidation. Oxidation is the loss of electrons.
Overall equation: $2H_2(g) + O_2(g) \rightarrow 2H_2O(g)$
Advantage: Only waste material from the reaction is water.
Disadvantage: Hydrogen is highly flammable so is difficult to handle and store safely.

6 a i $MnO_4^-(aq) + 8H^+(aq) + 5Fe^{2+}(aq) \rightarrow Mn^{2+}(aq) + 4H_2O(l) + 5Fe^{3+}(aq)$ **ii** The purple manganate(VII) ion changes to the very pale pink Mn^{2+} ion **b i** 2.34×10^{-5} moles **ii** 1.17×10^{-4} moles **iii** 0.026 g

7 a Copper carbonate: $[Cu(H_2O)_6]^{2+}(aq) + CO_3^{2-}(aq) \rightarrow CuCO_3(s) + 6H_2O(l)$ **b** Tetrachlorocuprate(II) $[Cu(H_2O)_6]^{2+}(aq) + 4Cl^-(aq) \rightleftharpoons [CuCl_4]^{2-}(aq) + 6H_2O(l)$

c 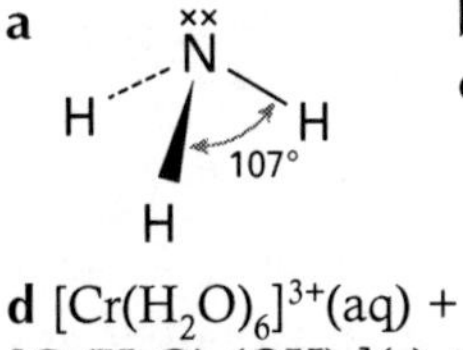

d i NaOH(aq) **ii** $[Cu(H_2O)_6]^{2+}(aq) + 2OH^-(aq) \rightleftharpoons [Cu(H_2O)_4(OH)_2](s) + 2H_2O(l)$ **e i** excess $NH_3(aq)$ **ii** ligand substitution reaction

8 a 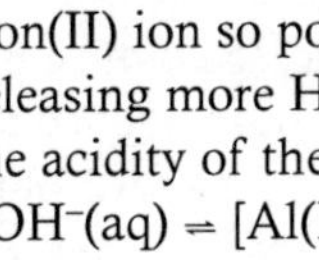

b H^+ acceptor **c** electron pair donor

d $[Cr(H_2O)_6]^{3+}(aq) + 3NH_3(aq) \rightleftharpoons [Cr(H_2O)_3(OH)_3](s) + 3NH_4^+(aq)$
$[Cr(H_2O)_6]^{3+}(aq) + 6NH_3(aq) \rightarrow [Cr(NH_3)_6]^{3+}(aq) + 6H_2O(l)$ **e** Reaction 2

9 a i $[Al(H_2O)_6]^{3+}(aq)$ **ii** $[Fe(H_2O)_6]^{2+}(aq)$ and $[Fe(H_2O)_6]^{3+}(aq)$ **iii** $[Fe(H_2O)_6]^{3+}(aq)$ has the lower pH as the iron(III) ion is more polarizing than the iron(II) ion so polarizes the water to a greater extent, releasing more $H^+(aq)$ ions and therefore increasing the acidity of the solution. **b** $[Al(H_2O)_6]^{3+}(aq) + 3OH^-(aq) \rightleftharpoons [Al(H_2O)(OH)_3](s) + 3H_2O(l)$

c Amphoteric substances react with both acids and bases: $[Al(H_2O)_3(OH)_3](s) + OH^-(aq) \rightleftharpoons [Al(OH)_4]^-(aq) + 3H_2O(l)$
$[Al(H_2O)_3(OH)_3](s) + 3H_3O^+(aq) \rightleftharpoons [Al(H_2O)_6]^{3+}(aq) + 3H_2O(l)$

10 **a** A catalyst is a substance that speeds up a chemical reaction by providing an alternative route of lower activation energy.
b

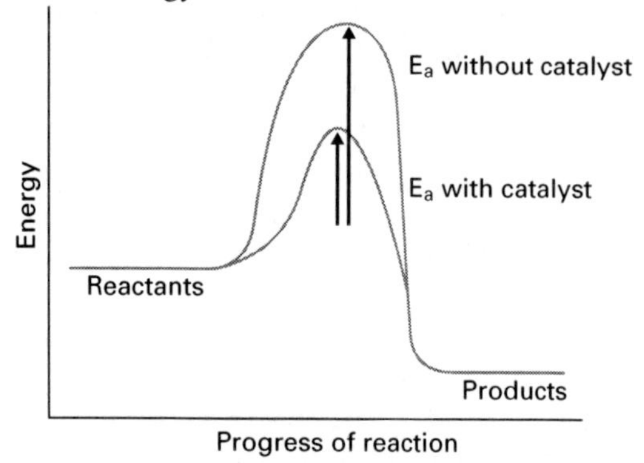

c A catalyst that is in the same state or phase as the reactants. **d** Bromine is the catalyst. **e** iron
f A catalyst support is an inexpensive solid such as a ceramic honeycomb which is coated with the catalyst. This provides a large surface area of catalyst from only a small quantity of material.

Answers to synoptic questions

Unit 4

1 **a** $C_5H_{10}O_2$ **b** $C_5H_{10}O_2$ **c** pentanoic acid **d** Add universal indicator and it would go orange/red; or react with a carbonate – bubbles of gas will be seen. If the gas is bubbled through limewater it turns cloudy. **e** weak – only partially dissociated. acid – proton donor **f** 2.72

2 **a** C_3H_8O **b** C_3H_8O propan-1-ol **c** **i** Orange solution turns green **ii** oxidation/redox **iii** propanal
d propanoic acid **e** propyl propanoate

3 **a** A is nitrobenzene, B is phenylamine **b** Benzene is highly toxic / repeated contact with benzene can cause leukaemia. **c** step 1 = nitration/electrophilic substitution step 2 = reduction
d **i** NO_2^+ nitronium ion
ii

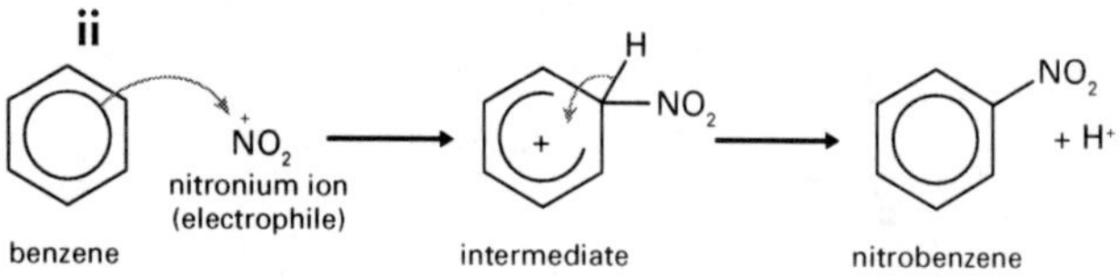

4 **a** Tollens' reagent – silver mirror/Fehling's solution – brick-red precipitate/ acidified potassium dichromate (VI) – orange to green **b** aqueous bromine – decolourises
c (*E*)-but-2-ene (*Z*)-but-2-ene

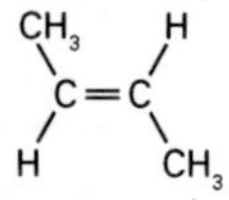

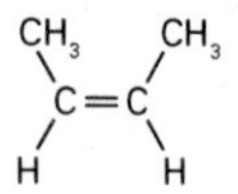

d For stereoisomers to exist a molecule must have: a C=C bond and two different groups attached to the carbon atoms involved in the double bond. But-1-ene has two hydrogen atoms attached to the carbon atoms involved in the double bond so does not meet the second requirement.

5 **a** **i** an ester **ii** propyl ethanoate
b

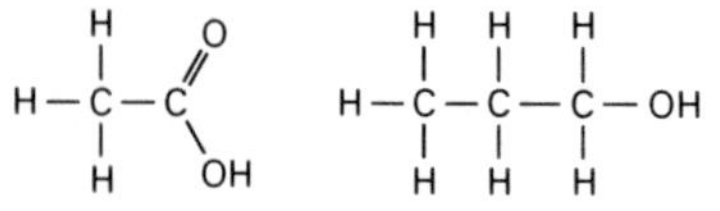

ethanoic acid propan-1-ol

c The ethanoic acid reacts with the alkali to form an ethanoate ion. **d** There will be two peaks. One will be three times the area of the other.

6 **a** It shows that the reaction is exothermic. **b** **i** time is less **ii** no effect on yield **c** **i** time is less **ii** yield decreases
d yield increases
e $K_c = \dfrac{[NH_3]^2}{[N_2]\,[H_2]}$ **f** 1.97 mol dm^{-3}

7 **a** **i** 22.5 g **ii** 15 dm^3 **b** 3– and 5+ **c** Lines drawn on the graph to show a constant half life. **d** rate = $k[N_2O]$

8 **a** **i** chloroethene (or vinyl chloride) **ii** addition polymerization **b** **i** methyl propanoic acid and ethanediol **ii** condensation polymerization
c polyester polyesters are hydrolysed in alkaline conditions.

9 **a**

haloalkane	colour of precipitate	identity of the precipitate
chlorooalkane	white	AgCl
bromoalkane	cream	AgBr
iodoalkane	yellow	AgI

b Silver bromide will not dissolve in dilute ammonia solution but it will dissolve in concentrated ammonia to form a colourless solution. **c** **i** CH_3^+, $C_2H_5^+$, $C_3H_7^+$
ii Chlorine has two isotopes; the 35 isotope gives a peak at 77 while the 37 isotope gives a peak at 79.

10 **a** reduction **b** +8 kJ mol^{-1}
c $K_a = [H^+]\,[CH_3CH_2COO^-] / [CH_3CH_2COOH]$
d **i** 2.79 **ii** Assumptions are that $[H^+] = [CH_3CH_2COO^-]$ and that $[CH_3CH_2COOH]$ = initial concentration of propanoic acid

Unit 5

1 **a** The enthalpy change when one mole (1) of a substance is formed from its elements in their standard states under standard conditions. (1)
b Elements in their standard states have a standard enthalpy of formation of zero. (1)
c $\Delta S^{\ominus} = \Delta S^{\ominus}$ (products) – $\Delta S^{\ominus}$ (reactants)
$\Delta S^{\ominus} = [(2 \times 28) + (3 \times 198)] - [51 + (3 \times 6)]$ (1)
$\Delta S^{\ominus} = 650 - 69 = 581$ J mol^{-1} (1)
$\Delta S^{\ominus} = 0.581$ kJ mol^{-1}(1)
$\Delta H^{\ominus} = \Sigma\Delta H_f^{\ominus}$ (products) – $\Sigma\Delta H_f^{\ominus}$ (reactants)
$\Delta H^{\ominus} = [(3 \times -111)] - [-1669]$ (1)
$\Delta H^{\ominus} = -333 + 1669 = 1336$ kJ mol^{-1} (1)
$\Delta G^{\ominus} = \Delta H^{\ominus} - T\Delta S^{\ominus}$
$\Delta G^{\ominus} = 1336 - (350 \times 0.581)$ (1)
$\Delta G^{\ominus} = 1132.65$ kJ mol^{-1} (1)
$\Delta G^{\ominus}$ is greater than zero so the reaction is not feasible. (1)
d **i** Aluminium is extracted by electrolysis. (1)
ii Aluminium oxide (1) dissolved in molten cryolite. (1)

2 a

substance	structure	bonds broken on melting	
sodium	giant	metallic bonds	(1)
silicon	giant	covalent bonds	(1)
phosphorus	simple	van der Waals forces	(1)

b $2Na(s) + 2H_2O(l) \rightarrow 2NaOH(aq) + H_2(g)$ (1) products, (1) balancing $pH = 12–14$ (1) $Cl_2(aq) + H_2O(l) \rightleftharpoons HCl(aq) + HClO(aq)$ (1) products, (1) balancing $pH = 1–3$ (1) **c** Amphoteric substances react with both acids and bases. (1) $[Al(H_2O)_3(OH)_3](s) + OH^-(aq) \rightleftharpoons [Al(OH)_4]^-(aq) + 3H_2O(l)$ (1) $[Al(H_2O)_3(OH)_3](s) + 3H_3O^+(aq) \rightleftharpoons [Al(H_2O)_6]^{3+}(aq) + 3H_2O(l)$ (1)

3 a i 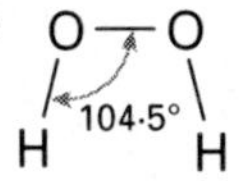

(2)

ii $1s^2\ 2s^2\ 2p^6\ 3s^2\ 3p^6\ 3d^3$ (1) Note that the electron configuration must be written *in full*. **iii** $[Cr(H_2O)_6]^{3+}(aq) + 3OH^-(aq) \rightarrow [Cr(H_2O)_3(OH)_3](s) + 3H_2O(l)$ reactants (1) products (1) **iv** $[Cr(OH)_6]^{3-}(aq)$ (1) **v** $2[Cr(OH)_6]^{3-}(aq) + 3H_2O_2(aq) \rightarrow 2CrO_4^{2-}(aq) + 2OH^-(aq) + 8H_2O(l)$ reactants (1), products (1) **b i** $Cr_2O_7^{2-}(aq) + 14H^+(aq) + 6Fe^{2+}(aq) \rightarrow 2Cr^{3+}(aq) + 7H_2O(l) + 6Fe^{3+}(aq)$ reactants (1), products (1) **ii** moles $Fe^{2+}(aq)$ reacting $= 1.25 \times 10^{-3}$ (1) moles $Cr_2O_7^{2-}(aq)$ reacting $= 2.08 \times 10^{-4}$ (1) volume of $Cr_2O_7^{2-}(aq)$ reacting $= 10.40\ cm^3$ (1) so, concentration $= 0.02\ mol\ dm^{-3}$

iii moles of $H^+(aq)$ needed $= 2.92 \times 10^{-3}$ (1) volume of $H_2SO_4(aq) = 14.59\ cm^3$ (1) Did you take account of the fact that one mole of $H_2SO_4(aq)$ releases two moles of $H^+(aq)$?

4 a i $Br_2(aq)$ (1) **ii** Mechanism is electrophilic addition (1)

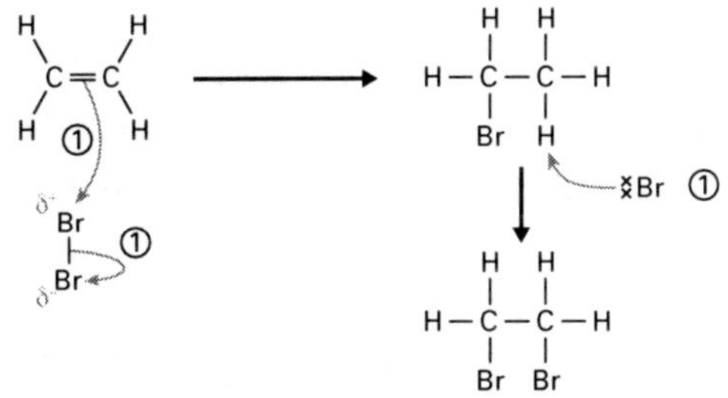

b i $BrCH_2CH_2Br + 2NH_3 \rightarrow H_2NCH_2CH_2NH_2 + 2HBr$ (1) **ii** Atom economy = (M_r of desired product $\div M_r$ of all products) $\times$ 100 (1) Atom economy = $(60 / 221.8) \times 100 = 27\%$ (1) Atom economy for reaction 1 = 100% (all additions are 100%) (1)
c A Lewis base is an electron pair donor. (1)
d i A bidentate ligand is able to donate two lone pairs of electrons. (1) **ii** Ethane-1, 2-diamine has two N atoms each of which has a lone pair that it is able to donate.

e 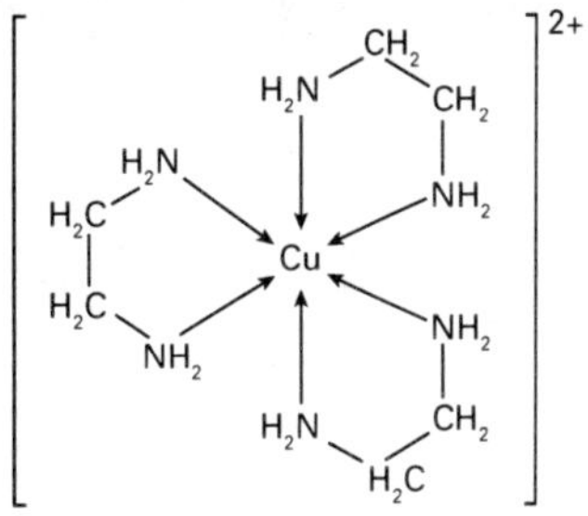

f The enthalpy change for ligand substitutions is almost zero as the same number and type of bonds are being broken and made (1) There is an increase in entropy as there are seven molecules on the right hand side and four molecules on the left hand side. (1) As a result the free energy change ΔG is negative and the reaction is feasible (1)

5 a Any two from: catalytic activity; variable oxidation states; the ability to form complexes; the ability to form coloured ions (2) Note – you will not get any credit for giving a property which is common to all metals, it must be one only shown by the transition metals. (1) **b i** Homogeneous catalysts are in the same state or phase as the reactants. (1)
$Fe^{2+}(aq)$ ions catalyse the reaction of $S_2O_8^{2-}(aq)$ and are regenerated in step two

Step 1 $S_2O_8^{2-}(aq) + 2Fe^{2+}(aq) \rightarrow 2SO_4^{2-}(aq) + 2Fe^{3+}(aq)$ (1)
Step 2 $2Fe^{3+}(aq) + 2I^-(aq) \rightarrow 2Fe^{2+}(aq) + I_2(aq)$ (1)
Overall $S_2O_8^{2-}(aq) + 2I^-(aq) \rightarrow 2SO_4^{2-}(aq) + I_2(aq)$

ii Heterogeneous catalysts are in a different state or phase to the reactants. (1) Vanadium(V) oxide catalyses the oxidation of $SO_2(g)$ in the contact process.

Step 1 $SO_2(g) + V_2O_5(s) \rightarrow SO_3(g) + V_2O_4(s)$ (1)
Step 2 $V_2O_4(s) + \frac{1}{2}O_2(g) \rightarrow V_2O_5(s)$ (1)
Overall $SO_2(g) + \frac{1}{2}O_2(g) \rightleftharpoons SO_3(g)$

6 a Iron forms two ions with partially filled d sub-levels and has its highest energy electron in a d sub-level. (1) Zinc only forms Zn^{2+} ions which do not have a partially full d sub-level but it does have its highest energy electron in a d sub-level (1) **b** $2Fe(s) + O_2(g) + 2H_2O(l) \rightarrow 2Fe^{2+}(aq) + 4OH^-(aq)$ reactants (1), products (1) **c i** The product in part **b** is formed as a green precipitate of $[Fe(H_2O)_4(OH)_2]$. (1). **ii** On standing in air a brown precipitate (1) of iron(III) hydroxide, $[Fe(H_2O)_3(OH)_3](s)$ is formed. (1)
d i A reducing agent is a species that brings about reduction. It is itself oxidized. (1) **ii** The electrode potential for Fe^{3+}/Fe^{2+} is more positive than that for Zn^{2+}/Zn which means that Fe^{3+} is more readily reduced. (1) The electrode potential for Fe^{3+}/Fe^{2+} is less positive than that for Cl_2/Cl^- which means that Cl_2 is more readily reduced than Fe^{3+}. (1)

The periodic table of the elements

Key

relative atomic mass **atomic number** name atomic (proton) number

1 (1)	2 (2)	(3)	(4)	(5)	(6)	(7)	(8)	(9)	(10)	(11)	(12)	3 (13)	4 (14)	5 (15)	6 (16)	7 (17)	0 (18)
							1.0 **H** hydrogen 1										4.0 **He** helium 2
6.9 **Li** lithium 3	9.0 **Be** beryllium 4											10.8 **B** boron 5	12.0 **C** carbon 6	14.0 **N** nitrogen 7	16.0 **O** oxygen 8	19.0 **F** fluorine 9	20.2 **Ne** neon 10
23.0 **Na** sodium 11	24.3 **Mg** magnesium 12											27.0 **Al** aluminium 13	28.1 **Si** silicon 14	31.0 **P** phosphorus 15	32.1 **S** sulphur 16	35.5 **Cl** chlorine 17	39.9 **Ar** argon 18
39.1 **K** potassium 19	40.1 **Ca** calcium 20	45.0 **Sc** scandium 21	47.9 **Ti** titanium 22	50.9 **V** vandium 23	52.0 **Cr** chromium 24	54.9 **Mn** manganese 25	55.8 **Fe** iron 26	58.9 **Co** cobalt 27	58.7 **Ni** nickel 28	63.5 **Cu** copper 29	65.4 **Zn** zinc 30	69.7 **Ga** gallium 31	72.6 **Ge** germanium 32	74.9 **As** arsenic 33	79.0 **Se** selenium 34	79.9 **Br** bromine 35	83.8 **Kr** krypton 36
85.5 **Rb** rubidium 37	87.6 **Sr** strontium 38	88.9 **Y** yttrium 39	91.2 **Zr** zirconium 40	92.9 **Nb** niobium 41	95.9 **Mo** molybdenum 42	98.9 **Tc** technetium 43	101.1 **Ru** ruthenium 44	102.9 **Rh** rhodium 45	106.4 **Pd** palladium 46	107.9 **Ag** silver 47	112.4 **Cd** cadmium 48	114.8 **In** indium 49	118.7 **Sn** tin 50	121.8 **Sb** antimony 51	127.6 **Te** tellurium 52	126.9 **I** iodine 53	131.3 **Xe** xenon 54
132.9 **Cs** caesium 55	137.3 **Ba** barium 56	138.9 **La *** lanthanum 57	178.5 **Hf** hafnium 72	180.9 **Ta** tantalum 73	183.9 **W** tungsten 74	186.2 **Re** rhenium 75	190.2 **Os** osmium 76	192.2 **Ir** iridium 77	195.1 **Pt** platinum 78	197.0 **Au** gold 79	200.6 **Hg** mercury 80	204.4 **Tl** thallium 81	207.2 **Pb** lead 82	209.0 **Bi** bismuth 83	210.0 **Po** polonium 84	210.0 **At** astatine 85	222.0 **Rn** radon 86
[223.0] **Fr** francium 87	[226.0] **Ra** radium 88	[227] **Ac†** actinium 89	[261] **Rf** rutherfordium 104	[262] **Db** dubnium 105	[266] **Sg** seaborgium 106	[264] **Bh** bohnium 107	[277] **Hs** hassium 108	[268] **Mt** meitnerium 109	[271] **Ds** darmstadium 110	[272] **Rg** roentgenium 111	Elements with atomic numbers 112-116 have been reported but not fully authenticad						

*** 58 – 71** Lanthanides	140.1 **Ce** cerium 58	140.9 **Pr** praseodymium 59	144.2 **Nd** neodymium 60	144.9 **Pm** promethium 61	150.4 **Sm** samarium 62	152.0 **Eu** europium 63	157.3 **Gd** gadolinium 64	158.9 **Tb** terbium 65	162.5 **Dy** dysprosium 66	164.9 **Ho** holmium 67	167.3 **Er** erbium 68	168.9 **Tm** thulium 69	173.0 **Yb** ytterbium 70	175.0 **Lu** lutetium 71
† 90 – 103 Actinides	232.0 **Th** thorium 90	231.0 **Pa** protactinium 91	238.0 **U** uranium 92	237.0 **Np** neptunium 93	239.1 **Pu** plutonium 94	243.1 **Am** americium 95	247.1 **Cm** curium 96	247.1 **Bk** berkelium 97	252.1 **Cf** californium 98	[252] **Es** einsteinium 99	[257] **Fm** fermium 100	[258] **Md** mendelevium 101	[259] **No** nobelium 102	[260] **Lr** lawrencium 103

INDEX